P/B -Thorsons, first ed. 1984   OS/ 9071 /health
very light tanning - o/w G.C. 112 pp inc
                                        index

Evening Primrose Oil

£ 1.25

Its remarkable properties and its use in the treatment of a wide
range of conditions.

KT-161-998

By the same author
MULTIPLE SCLEROSIS — A Self-help Guide to Its
Management
THE Z FACTOR — How Zinc is Vital to Your Health
(with Dr Michel Odent)

Evening Primrose Oil

by
JUDY GRAHAM

THORSONS PUBLISHING GROUP
Wellingborough, Northamptonshire
·
Rochester, Vermont

First published March 1984

© JUDY GRAHAM 1984

10   9

All rights reserved. No part of this book may be reproduced or utilized in any form or by any means, electronic or mechanical, including photocopying, recording or by any information storage and retrieval system, without permission in writing from the Publisher.

British Library Cataloguing in Publication Data

Graham, Judy
Evening Primrose Oil.
1. Evening Primrose Oil — Therapeutic use
I. Title
615'.323'672   RS165.B/

ISBN 0-7225-0743-7

Printed in Great Britain by Woolnough Bookbinding,
Irthlingborough, Northamptonshire

This book is dedicated to the doctors who are working to solve the many diseases of civilization.

'Only connect.'
E.M. Forster

The Evening Primrose
by John Clare

When once the sun shrinks in the west,
And dew-drops pearl the evening's breast;
Almost as pale as moonbeams are,
Or its companiable star,
The evening primrose opens anew
Its delicate blossoms to the dew
And, hermit-like, shunning the light,
Wastes its fair bloom upon the night;
Who, blind-fold to its fond caresses,
Knows not the beauty he possesses.

Thus it blooms on while night is by;
When day looks out with open eye,
'bashed at the gaze it cannot shun,
It faints and withers and is gone.

CONTENTS

ACKNOWLEDGEMENTS

My thanks are due to the many doctors who have done the research work on evening primrose oil. I have freely quoted from their papers in order to keep as closely as possible to their own accounts of their findings. The references for each of these papers can be found at the back of the book.

My acknowledgements to Dr Hugh Carmichael, consultant physician, Vale of Leven District General Hospital, Dunbartonshire; Dr John Rotrosen, assistant professor of psychiatry, New York University School of Medicine; Professor Robert Zurier, Chief, Rheumatology Section, University of Pennsylvania School of Medicine; Dr Paul Preece, consultant surgeon, Ninewells Hospital, Dundee; Dr Michael Brush, Department of Gynaecology, St Thomas' Hospital, London; Drs Steven Wright and John Burton, Department of Dermatology, Bristol Royal Infirmary, Bristol; Dr Allan Campbell, consultant physician, Hairmyres Hospital, East Kilbride; Dr G.C. MacEwan, consultant opthalmologist, Gartnevel General Hospital, Glasgow; Dr M.A. Mir, consultant physician, University Hospital of Wales, Cardiff; Dr K.S. Vaddadi, consultant physician, Bootham Park Hospital, York; Dr I. Glen, consultant psychiatrist, Craig Dunain Hospital, Inverness.

I would also like to thank Valerie Grundon, Howard Thomas, Peter Lapinskas, David Dow, Leslie Smith and John Williams for their special help.

Particular thanks are due to Dr David F. Horrobin, Director of the Efamol Institute in Nova Scotia, Canada, and to his co-workers. And I owe a personal thankyou to Dr Michel Odent for his support and encouragement while I was writing the book.

PREFACE

It might seem very odd that someone who is neither a scientist nor a botanist should be writing a book about evening primrose oil. I even find it quite surprising myself that I have got so involved in this bizarre little plant and its possibilities.

My own involvement with evening primrose oil goes back to the early 1970s. It was then that I began taking it (for multiple sclerosis) and have carried on ever since.

In my book *Multiple Sclerosis — A Self Help Guide to Its Management* (Thorsons, 1981) I devoted a whole chapter to evening primrose oil, and this book came to be written as a development of that chapter.

I also became fascinated by all the marvellous possibilites of evening primrose oil, and followed the various scientific studies with great interest.

Even though I take the oil myself, this book has been written with the objectivity required of any journalist. I am simply reporting the facts and letting them speak for themselves.

INTRODUCTION

The story of the evening primrose could turn out to be one of the most fantastic in plant and medical history. Like foxglove (digitalis), cinchona bark (quinine) and rauwolfia (reserpine) before it, the evening primrose promises to take its place in the hall of fame of plants with important medicinal properties. But unlike these other natural products which are, on the whole, useful for only one condition, the oil of the evening primrose has properties which make it useful for a very wide range of illnesses.

A sceptic might think he'd strayed into a Victorian fairgound where some charlatan was selling strange mixtures promising to cure everything from gout to hiccups with one gulp. And, on the face of it, the claims made for evening primrose oil do seem quite fantastic. After all, how can the same unassuming little plant be used in the treatment of benign breast disease *and* brittle nails *and* faulty blood vessels, and a score of other ailments?

The answer is that the mechanism for each of these conditions is similar. Put simply, the oil of the evening primrose converts, in the body, to a physiologically-active substance called Prostaglandin E1 (PGE1), which is the real hero of the piece. It's the PGE1 which is doing much of the good work in each case, although making prostaglandins is not the only talent of the oil — its fatty acid provides for cell membrane growth as well.

The unique quality of evening primrose oil is that it contains a substance called gammalinolenic acid (GLA), and it is this which eventually converts in the body to PGE1 (see page 26). The evening primrose is the only plant to have this vital ingredient with no toxic properties, though the sea alga, spirulina, does also contain gammalinolenic acid. So does borage, anchusa, rose bay willow herb, comfrey and some

mosses, but none of these is readily available for human
consumption. Human breast milk, which has countless
nutrients essential for a baby, also contains GLA. But now
evening primrose oil is being called 'mother's milk for adults'.

Evening primrose oil capsules are now being sold over the
counter in health food shops and chemists as a dietary
supplement of essential fatty acids. But at universities and
hospitals in England, Scotland, Wales, the U.S.A., Canada,
Australia and South Africa, strictly controlled research trials are
underway to put evening primrose oil to the test. It is virtually
unheard of for a health food product to be submitted to such
stringent scientific study, which gives you some indication of
the importance of the evening primrose.

By the time you read this book, evening primrose oil may
well have the status of a pharmaceutical and be available on
prescription in Britain on the National Health Service, for one
or more of the conditions listed in this book. When this
happens, it will be one of the few pharmaceuticals available
which are not made from synthetic chemicals, but from entirely
natural products.

1.   WHAT IS THE EVENING PRIMROSE?

Most people have never heard of the evening primrose, and not surprisingly. It's not a well known plant, and certainly not one you're likely to grow in your back garden, although it is rather a pretty botanical specimen. Mostly it grows wild along roadsides, railway sidings and on waste sites, and it thrives in places such as sand dunes.

Strictly speaking, the evening primrose is not a primrose at all. It's not even related to the primrose family. It belongs to the willow herb family, and in its wild state is more a weed than a plant.

It has acquired its name because its bright yellow flowers resemble the colour of real primroses, and because its flowers open in the evening. The flower usually lasts for the whole of the next day, particularly in dull weather, but in bright sunlight the flowers fade quite quickly. In England, the plant flowers from the end of June to mid August.

Experts who classify plants (taxonomists) will tell you that the evening primrose belongs to the order Myrtiflorae, family Onagraceae, genus *Oenotherae*. The generic name comes from the Greek *oinos* (wine) and *thera* (hunt). According to herbals, this described a plant — probably a willow herb — which gave one a relish for wine if the roots were eaten. Another interpretation is that the plant dispelled the ill effects of wine, and this fits in better with modern research (see Chapter 17).

The French are supposed at one time to have given it the romantic name of '*Belle de Nuit*'. Nowadays, it's known there as '*L'herbe aux ânes*', though its connection with donkeys is unclear! More of a mouthful is the name given to it by present-day seed merchants and horticulturalists. They call one type *Oenothera biennis*, and another type *Oenothera lamarkiana*. These are the two types in use today.

The evening primrose plant and its seeds have been used by

American Indians for hundreds of years.[1] According to folklore, a tribe called Flambeau Ojibwe were the first to realise the medicinal properties of the evening primrose plant. They used to soak the whole plant in warm water to make a poultice to heal bruises, and they also used the plant for skin problems and asthma.

Herbals describe the evening primrose as being astringent and sedative, and the oil helpful in treating gastro-intestinal disorders, asthma, whooping cough, female complaints, and wound healing.

History

The evening primrose originated in North America, and botanists first brought the plant from Virginia to Europe in 1614 as a botanical curiosity. Most of the strains, however, came to Britain as 'stowaways' in cargo ships carrying cotton. As cotton is light, soil was used as ballast. The ballast was dumped on reaching port, and with it were dumped stray seeds of evening primrose. Even today there are areas around the major ports, such as Liverpool, where evening primrose plants — descendants of the cotton ballast — grow in profusion.

In Europe, the evening primrose became known as 'King's Cure All' by those who knew its almost magical medicinal properties. For centuries, however, the evening primrose was left to straggle along without anyone taking much notice. It wasn't until this century that scientists began to look at the plant for its industrial potential in such things as paint.

In 1917 a German scientist called Unger examined the plant, and found that the seeds contained 15% oil, which was extractable with light petroleum.[2] In 1919 the Archives of Pharmacology published a paper by Heiduschka and Luft who were the first to do a detailed analysis of the oil. They extracted 14% oil with ether, and apart from the normal oleic and linoleic acids, found a new fatty acid, which they named γ-linolenic acid (gammalinolenic acid).[3] In 1927, three German scientists repeated the Heidushka and Luft test, and came up with a more detailed analysis of the chemical structure of this gammalinolenic acid.[4]

Twenty-two years later Dr J.P. Riley, a British biochemist in the Department of Industrial Chemistry at Liverpool

University, came across the German papers on evening primrose oil and decided to analyse the oil for himself, but this time using modern techniques. So Dr Riley set off for the sandhills near Southport in Lancashire and picked a bunch or two of evening primrose plants. He dried the plants, separated the seeds, and extracted the oil. To his great satisfaction, he found for himself the unique gammalinolenic acid.[5]

It wasn't until the 1960s, however, that British scientists began investigating the oil for its possible health uses. The first experiment was on rats. The aim of this experiment was to compare the biological activity of the commonly-found linoleic acid with the rare gammalinolenic acid (GLA).[11]

The rats were put on a diet lacking in essential fatty acids, and after a few weeks they developed loss of hair and skin problems. They were then divided into two groups. One group was fed linoleic acid and the other group was fed gammalinolenic acid. The results of this first experiment were remarkable. The rats in the GLA group recovered more rapidly than the other group, and there was evidence that the GLA was far more efficiently taken up by the cells of all the important tissues and organs of the body.

The success of this experiment encouraged biochemists and other scientists to do more research work on this unusual seed oil. The next test was to investigate the effect of the GLA in regulating cholesterol production in a group of rabbits being fed a diet high in animal fats and to compare the results with a control group being fed a normal diet. The results showed that the GLA in evening primrose oil could control blood cholesterol levels. Further experiments followed, opening up yet more possibilities, particularly in heart disease research.

Much of this research during the 1960s was the brainchild of biochemist John Williams. If it had not been for this one man, evening primrose oil might well have been lost in the archives of obscure journals forever. When a large pharmaceutical company bought the small firm where John Williams had been doing all the work on evening primrose oil, their policy not to get involved with natural products meant they decided to drop the seed oil project altogether.

John Williams took an early retirement, but during the year after he left the company he couldn't get the evening primrose oil project off his mind. Fortunately, with the help of a Cheshire businessman, John Williams was able to take over the project and buy all the seed stock, documents and patents and start up his own company, Bio Oil Research Ltd. They began to manufacture capsules of evening primrose oil under the brand name *Naudicelle* and carried on with their scientific studies into the nutritional benefits of the oil.

During the 1970s, many other scientists took a keen interest in the possibilities of evening primrose oil. Drs Ahmed Hassam and Michael Crawford of the Nuffield Laboratories of Comparative Medicine in Regents Park, London, did various experiments which showed that GLA was ten times more biologically effective than linoleic acid.[6,7]

The first disease for which evening primrose oil was used was Multiple Sclerosis (see Chapter 13). This followed the publication of a paper by some eminent neurologists which said that sunflower seed oil reduced the severity and frequency of relapses in the disease.[8] Dr Hassam's evidence suggested that if sunflower seed oil helped a little, then evening primrose oil, being that much more active, could help even more.

A leading specialist in MS, Professor E.J. Field of Newcastle University, England, tested the oil and did find an increased activity compared to ordinary sunflower seed oil.[9] On Professor Field's advice, many people with MS have been taking evening primrose oil capsules ever since these findings in 1974, even though there have not been conclusive enough results in controlled trials for the oil to be made available on the British National Health Service.

A neurophysiologist who at one time worked in the same university as Professor Field was responsible for catapulting evening primrose oil into the wider fields of international medical research. Dr David Horrobin, at that time Reader in Physiology at Newcastle University Medical School, first became interested in evening primrose oil in the mid-1970s, as a result of his work on schizophrenia and prostaglandins.

Dr Horrobin, with Dr Abdulla of Guy's Hospital, London, discovered that schizophrenics have extremely low levels of Prostaglandin E1 (PGE1), which is manufactured in the body from gammalinolenic acid.[10] That finding also set Dr Horrobin thinking about the role of PGE1 in many other diseases. This enterprising young doctor was so convinced of the far-reaching possibilities of gammalinolenic acid that he persuaded a director of Agricultural Holdings Ltd., (which included the seed merchants who grew evening primroses, Hurst Gunson Cooper Taber Ltd.) to set up another company to market and develop evening primrose oil.

So Efamol Ltd., was founded, and the product *Efamol* launched. Dr Horrobin is now Medical Director of Efamol Ltd., in the U.K., and President and Research Director of Efamol Research Inc., in Novia Scotia, Canada. In this new position, Dr Horrobin made contact with the doctors who are doing most of the research trials, and helped draw up the protocols.

This brings the history of the evening primrose up to the present day, when there are over 100 clinical trials in progress, and several companies around the world marketing evening primrose oil.

2. A UNIQUE BOTANICAL SPECIMEN

Botanists first became interested in the evening primrose plant at the turn of the century when it was thought to be breaking the laws of inheritance discovered a few years earlier by Mendel.

In ordinary species, if two plants are cross-pollinated, the first generation of offspring are identical to one another. But in the second generation they form a mixture of two types, intermediate between the two parents. In further generations, new variants continue to be produced.

The evening primrose, however, does something very different. If two plants are crossed, the first generation usually turns up as a mixture of two types which don't resemble either of the two parents or each other. In the next generation, the two groups do not split up into a mixture. Instead, the plants breed true. This can be repeated for many generations without any variation.

Within the main species of the evening primrose, there are countless different varieties — at least 1000. This is one of the consequences of the plant's peculiar genetic system — there can be an astronomic number of different types of plant, each of which breeds true.[1]

From a distance, a cluster of plants might look the same. But at a closer look one sees there are some subtle variations between one type and another. The size of the flowers can vary from $\frac{1}{2}$ to 5 in. (1 to 13 cm), and the pigment may differ in each plant. Although most flowers are coloured yellow, some have a mauve hue, and there are many variations in colour in each part of the plant. There is a huge difference in plant size, too. Some grow no higher than dandelions, while others can shoot up to more than 8 feet (2.4 metres) high.

Some evening primrose plants need bees or moths for pollination; others are self-pollinated. The variations provide interesting opportunities for plant breeders.

From wild flower to crop
In its natural state, the evening primrose is a wild flower. The interesting thing is that agronomists are only now turning this wild, untamed flower into a crop. Most crops have been in existence for centuries and farmers have had plenty of time to learn about their cultivation. The evening primrose, however, is a brand new crop and presents an enormous challenge to agronomists who are trying to master its cultivation in a fraction of the time devoted to other crops. Certainly, the evening primrose is by no means the only ornamental plant to be grown as a crop, but it is the newest.

The breeding programme
A great deal of research effort is being put into breeding better varieties by experts like Dr Peter Lapinskas, plant breeder at the Hurst Crop Research and Development Unit at Great Domsey Farm in Essex, England.

His main breeding objectives are to produce varieties which will give the highest possible yield of oil with the highest levels of GLA, with the greatest reliability and at the lowest cost.

Just as the evening primrose plant can vary so much in its characteristics, so the oil content and quality can differ considerably from one type to another. The plants used at the moment yield 7 - 9% GLA. But the potential GLA yield of the evening primrose is probably more than double that.

Making new hybrids
A plant breeder selects the plants which have the best combination of characteristics he is looking for in the ideal plant. For instance, one plant might be chosen for characteristic A, and another for B. He makes crosses between them and selects plants in the progeny which combine both A and B.

Cross-pollination is a time-consuming process. Firstly, the anthers from the flowers of the chosen female parent are removed, and the emasculated flowers protected from pollen-carrying insects. The next day, pollen from the male parent is applied to the receptive female part of the flower (the stigma). Once the cross has been made, the flower is labelled to identify it for when it is later harvested. To ensure success, dozens of flowers are pollinated in this way for each cross which is required. Even though plant breeders at research farms are still trying to create the very best plant, the type already being grown is perfectly good enough in the quality and quantity of its oil yield, and the *Oenothera Biennis* and *Lamarkiana* are being grown in countries as diverse as Spain, Britain, Hungary and the U.S.A.

Growing the best crop

The aim of any agronomist is to produce a strain in the most efficient and cost-effective way so that the farmer will get a reliable crop with the maximum yield and of the highest quality which will provide the consumer with the best and lowest-priced product. With an entirely new crop, a wild plant which is still being tamed, there are many unanswered questions. When, for instance, do you sow the seed? How deep should you sow the seed? What are the moisture requirements? An agronomist also needs to find out how far apart the plants should be spaced. Should they be in rows? Do they need fertilizer? Are there particular weed problems? Do they suffer from any diseases or troubled by particular pests? And what about the type of soil? And when is the best time to harvest?

So far, the agronomists working on the evening primrose plant have come up with many of the answers. They've also found out that it is very adaptable and seems to be able to grow in virtually any type of soil. Another bonus with the evening primrose is that its multiplication rate is very high. One kilo of seed multiplies into 1000 kilos for the farmer.

Quality control

A uniform seed is used, and farmers are advised on how to get the best results from the evening primrose crop. Batches of seed are routinely analyzed to make sure no contaminants have got in, and that the GLA content conforms to the specification.

The oil is also routinely analyzed by a machine called a Gas Liquid Chromatograph (GLC). This separates out the fatty acids — palmitic, oleic, stearic, linoleic, and gammalinolenic — and enables one to estimate the proportion of each fatty acid from a microprocessor print-out. The GLC machine has revolutionized research into essential fatty acids with its easy and accurate analysis.

Similarly stringent checking goes on during the extraction and encapsulation processes, to make sure of a pure high quality yield with exactly the same amount of oil in each capsule.

3. ESSENTIAL FATTY ACIDS

There was a time when no one knew what Vitamin C was, or why shiploads of sailors were dying of scurvy. Perhaps we are in the same boat today! Few people know what essential fatty acids are, yet they could soon prove to be as important, and as well known, as vitamins, minerals or proteins are today.

What are essential fatty acids?

Firstly, it might help to say what essential fatty acids are *not*, because 'fatty' is a rather misleading word and makes you think of fatty things like pork dripping or cream buns. However, essential fatty acids have nothing whatsoever to do with that kind of fat. Think of them more like protein, or vitamins. Indeed, essential fatty acids are sometimes called Vitamin F. And this is really quite a sensible name for them because EFAs are vitamin-like lipids. They are called 'essential' because the body must have them and can't make them by itself, so you have to eat foods that contain them — foods like sunflower seed oil, safflower seed oil, corn oil, liver, kidneys, brains, sweetbreads, lean meat, legumes (these all belong to the linoleic acid family), and green vegetables, fish, shellfish, fish liver oils and linseeds (these belong to the alpha-linolenic acid family). Evening primrose oil is very rich in essential fatty acids, and belongs to the linoleic acid family.

The importance of essential fatty acids

Fatty acids are an essential part of nutrition and they perform all kinds of vital functions within the body: they give energy; they help to maintain body temperature; they insulate the nerves; they cushion and protect the tissues. They are part of the structure of every cell in your body and are vital for metabolism. They are also the precursors of the all-important short-lived regulating molecules, the prostaglandins[3] (see next chapter).

The Food and Agricultural Organization and the World Health Organization suggest that a minimum of 3% of total calories should be from essential fatty acids in adults, and 5% in children and pregnant and breastfeeding women.[4] Roughly 60% of the brain is made up of lipids, of which an important part is the essential fatty acids. They are vital for the proper growth and development of the brain and the central nervous system.

Essential fatty acids play a fundamental role in all cell membranes of the body.[5] How fluid and flexible the cell membranes are depends on how much essential fatty acids they have. And this has a big influence on the lymphocytes, the white blood cells, who need to be strong and healthy as their job is to rid the body of foreign invaders. The activity of these lymphocytes may depend on the state of the cell membrane, and they will behave differently according to whether the cell membrane is fluid (plenty of EFAs), or rigid (not enough EFAs).

EFA deficiency

Experiments done on animals who were deprived of essential fatty acids, showed a variety of unpleasant symptoms. These included:

- Abnormalities of the heart and circulation.
- Poor skin.
- Wounds did not heal properly.
- Failure to reproduce (especially in males).
- Inflammatory disorders and arthritis.
- Failure of normal brain development.
- Dried up tear ducts and salivary glands.
- Faulty immune function.

The earliest results of such experiments were in 1929.[23]

There is, of course, no shortage of foods rich in linoleic acid and they are all readily available. So how is it possible to be short of essential fatty acids? The problem is that the essential fatty acids in the food you are eating may not be getting through to the places they're needed most because of some metabolic defect.

The Metabolic Pathway

The metabolic pathway from linoleic acid is a bit like an obstacle race, with lots of hurdles between the start and finish. With many people (quite possibly the ones suffering from the conditions in this book), the obstructions blocking the metabolic route are so great that the journey never gets finished.

Linoleic acid on its own has no biological activity. To be of any use whatsoever it has to be converted in the body to other things which are biologically active. On a clear run, the metabolic pathway looks like this:

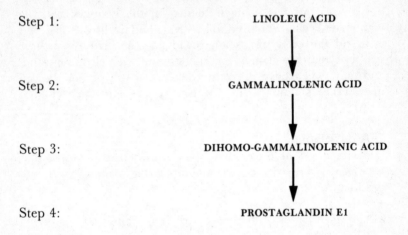

Step 1:   LINOLEIC ACID

Step 2:   GAMMALINOLENIC ACID

Step 3:   DIHOMO-GAMMALINOLENIC ACID

Step 4:   PROSTAGLANDIN E1

Figure 1. The Metabolic Pathway of Linoleic Acid.

Whereas on a bumpy run, all kinds of nasties can get in the way, which we'll call 'blocking agents.'

Blocking agents

The blocking agents get in the way of the conversion from one step along the metabolic pathway to the next. Their favourite place for setting up a road block is between linoleic acid and gammalinolenic acid, steps 1 and 2.

These are the most common blocking agents:

- Foods rich in saturated fats.
- Foods rich in cholesterol.

- Foods rich in trans fatty acids (see below).
- Alcohol, in moderate to large amounts.
- Diabetes (lack of insulin).
- Ageing.
- Viral infections.
- Radiation
- Cancer.
- Zinc deficiencies.

With so many blocking agents[6,7,8,9] it's not surprising that people living in the western world may not be metabolizing essential fatty acids, even though they may be eating plenty of them.

A few of the blocking agents are unavoidable. But you have a fair amount of control over most of them, particularly too much drink, and foods high in cholesterol and saturated fat. But what about these trans fatty acids? What are they, and how do you know when you're eating them?

Trans fatty acids

The only type of linoleic acid which can convert into biologically useful things is what is known as cis-linoleic acid. This is when the oil (e.g. corn oil, sunflower seed oil) is in its natural, unadulterated state.

Only cis-linoleic acid has any real value as an essential fatty acid. Luckily, evening primrose oil contains over 70% cis-linoleic acid, as well as about 9% cis-gammalinolenic acid, and none of this is lost when the oil is put into capsules.

Once linoleic acid gets processed and hydrogenated, it turns into a biologically different form and becomes a real non-starter as far as the metabolic pathway is concerned. This is because what were originally biologically active essential fatty acids have been turned into biologically inactive trans fatty acids.

You may never have heard of them, but trans fatty acids are false friends lurking in our everyday foodstuffs. On the face of it, they look perfectly pleasant and often delicious: ordinary bottles of cooking oil on the supermarket shelves; cartons of margarine; luscious looking pastries; lovely sweets; tasty french fries; and all those kinds of foods which people generally love to eat. But beware! Trans fatty acids behave as if they were saturated fats. And far from being essential fatty acids, they actually produce EFA deficiency states and compete with genuine EFAs for your body's time and attention. They elbow the real EFAs out of the action.[10,11,12,13,22]

The trans fatty acids which you eat make their way into tissues like the brain, heart and lungs, and some scientists are sure that they change the properties of these tissues — probably for the worse.[16,17,18]

Here is a table from the United States which shows the percentages of trans fatty acids in foodstuffs:[19]

- Bakery products — up to 38.5.%
- Sweets — up to 38.6%
- French fries — up to 37.4%
- Hard margarines — up to 36%
- Soft margarines — up to 21.3%
- Diet margarines — up to 17.9%
- Vegetable oils — up to 13.7%
- Vegetable oil cooking fats — up to 37.3%

The cockeyed thing is that when government and other bodies involved in the nation's nutrition examine the amount of lipids (fats) people are eating, they lump together both trans and cis fatty acids,[19,20,21,10] perhaps not being fully aware that trans fatty acids may as well not be counted for all the nutritional good they do. This means that the overall intake of real essential fatty acids is much lower than we have been lead to believe.

It's interesting to ponder the fact that it's only since the 1920s that significant amounts of trans fatty acids began to be added to the diet, (though they have always existed in small amounts in dairy produce). People who are interested in the geographical distribution and the increase in western diseases might find this worthy of more research.

Trans fatty acids are by no means the only blocking agents, but they are the ones which cause innocent eaters to fall into the trap of thinking they're eating the right things when they're not. So you need to look out for those in particular.

As if all these obstacles weren't enough, the cis-linoleic acid, which is trying to do its best to make all the steps to convert to Prostaglandin E1, needs some friendly helpers along the way. And if these helpers aren't around, the going gets tough.

STEP 1: CIS-LINOLEIC ACID

enzyme delta-6-desaturase
needed to get to step 2

helped by:
zinc, magnesium, Vitamin B$_6$, biotin

BLOCKED BY TRANS FATTY ACIDS
BLOCKED BY SATURATED FATS
BLOCKED BY CHOLESTEROL
BLOCKED BY TOO LITTLE ZINC
BLOCKED BY TOO LITTLE INSULIN
BLOCKED BY TOO MUCH ALCOHOL
BLOCKED BY AGEING
BLOCKED BY CERTAIN VIRUSES
BLOCKED BY CHEMICAL CARCINOGENS
BLOCKED BY IONIZING RADIATION

STEP 2: GAMMALINOLENIC ACID
(Evening Primrose oil starts here)

STEP 3: DIHOMO-GAMMALINOLENIC ACID

helped by:
Vitamin C, Vitamin B$_3$

STEP 4: PROSTAGLANDIN E1

Figure 2. The Bumpy Metabolic Road of Cis-linoleic Acid (simplified)

Note: To prevent oxidation of the oil inside the body, evening primrose oil *must* be taken with vitamin E.

Co-factors

Scientist sometimes call these helpers 'co-factors'. During the important journey between cis-linoleic acid and gammalino-lenic acid, the help of an enzyme called delta-6-desaturase is needed. If this enzyme isn't there or is not doing its job very well it can cause all sorts of havoc, as you'll see in later chapters. As well as enzymes, the essential fatty acid travellers need vitamins and minerals as aides along the way, particularly Vitamin B_6, Vitamin C, Vitamin B_3, biotin, zinc, and magnesium. A real shortage of any of these can mean trouble.

The importance of GLA

The exceptional thing about evening primrose oil is that it has an enormous headstart in the metabolic obstacle race. In fact, it has a clear and unobstructed run because it starts life at step 2!

The active ingredient of evening primrose oil is gammalinolenic acid (GLA). This means that all the blocking agents lying in wait between steps 1 and 2 are of absolutely no consequence. Evening primrose oil by-passes them completely. Its clear run looks something like this:

Step 2:   **GAMMALINOLENIC ACID**
Step 3:   **DIHOMO–GAMMALINOLENIC ACID**
Step 4:   **PROSTAGLANDIN E1**

See also Figure 2 (page 29).

4. PROSTAGLANDINS

This book is called evening primrose oil, but perhaps the name that should be up in lights is somebody else's because the real star is Prostaglandin E1. Evening primrose oil only deserves credit because it's what's called a 'precursor' of Prostaglandin E1. And it's a much more reliable precursor of PGE1 than linoleic acid, which, as we saw in the last chapter, can get waylaid on its metabolic pathway.

PGE1 is something of a little miracle worker. It's one of the unsung heroes of modern medicine. Most people who aren't doctors or scientists have barely even heard of prostaglandins, except perhaps vaguely as something to do with childbirth. Yet the discovery of their importance might turn out to be crucial in the history of medicine.

This is what Dr David Horrobin (who has done much of the frontier-breaking research work) has to say about the revolutionary possibilities of prostaglandins:

> I believe that within the next decade we in the prostaglandin field have the opportunity to bring about a revolution in both biology and practical medicine which will have a greater impact on people's lives and on fundamental biological concepts than any previous bio-medical revolution. Few can doubt that the prostaglandins will have an impact approaching that to the antibiotics 30 years ago.[13]

Prostaglandins, despite being unknown by ordinary people, are fast becoming the biggest growth area in biochemistry. I bet it won't be long before the word 'prostaglandin' gets bandied about as the 'in' thing and becomes the newest medical breakthrough.

What are prostaglandins?

It is only because of a scientist's mistake that prostaglandins are called by their mouthful of a name at all. They were discovered by a Swedish scientist called von Euler in the 1930s. He first found these new molecules in the seminal fluid and thought they came from the prostate gland, so he called them prostaglandins.

What they do is act as vital regulators. They control every cell and every organ in your body on a second-by-second basis. The nearest thing to them is hormones, which also have important messenger roles. But prostaglandins aren't like hormones, which zip around all over the place. Prostaglandins are much more local than that. So they're a bit like friendly neighbourhood hormones, regulating everything only on their home patch.

Each prostaglandin has a very specific effect in each tissue. Generally, prostaglandins help to control what each and every cell is doing, and they regulate the activity of certain key enzymes. This vital job as a cell regulator helps explain why the unsung hero of this book, PGE1, keeps popping up in every chapter. And because the prostaglandins have such a key role as regulators of every cell, it helps to explain why this important molecule could possibly be able to be of such vital importance in so many apparently different conditions.

Since von Euler's discovery in the 1930s, scientists have located prostaglandins pretty well everywhere, though medical papers highlight the finds in blood vessel walls;[1] macrophages;[2] platelets;[3] duodenal secretions;[4] nerves;[5] and every organ.

Prostaglandins have an extremely short life span. In the time it's taken you to read this sentence they've done their job and clocked off. Maybe they're so short-lived because they are naturally unstable. Or perhaps it's because they have highly efficient mechanisms which break them down. Either way, most PGs are removed from the blood during a single passage through the lungs.

Types of prostaglandins

It's as well at this point to give a brief mention to the various relatives of PGE1, as some of them will be cropping up again later. There are three series of prostaglandins; PG1, PG2, and PG3. Each of these has a different chemical structure. Just to make things slightly more complicated, within each series there are many types of PGs, such as A,B,D,E,F etc. In all, there are at least 30 prostaglandins.

In man, the three series of prostaglandins is each derived from a different fatty acid. Series 1 and 2 both come from the linoleic acid family. Series 3 PGs are from eicosapentaenoic acid, a member of the alpha-linolenic family and most commonly found in oily sea foods.

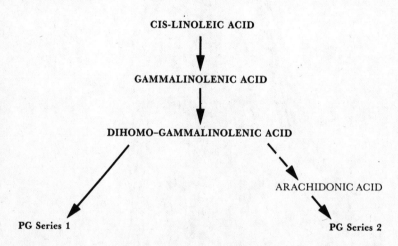

CIS-LINOLEIC ACID

GAMMALINOLENIC ACID

DIHOMO–GAMMALINOLENIC ACID

ARACHIDONIC ACID

PG Series 1

PG Series 2

Figure 3. The Precursors to Prostaglandin Series 1 and 2.

Evening primrose oil, whose active ingredient is gammalinolenic acid, is a precursor of the Series 1 PGs.

Good and bad prostaglandins

We already know that the 'Superprostaglandin' is PGE1. But amongst its relatives there's a fair share of 'goodies' and 'baddies'. In fact, some prostaglandins are considered such nuisances that scientists have come up with prostaglandin-blocking drugs for putting them out of action. (Aspirin works along these lines, though for decades scientists didn't know how it worked.) Some of the recent research on prostaglandins has won the Nobel Prize for the scientists concerned.

The bad prostaglandins are notorious trouble-makers and are usually on the scene of crime when there's inflammmation

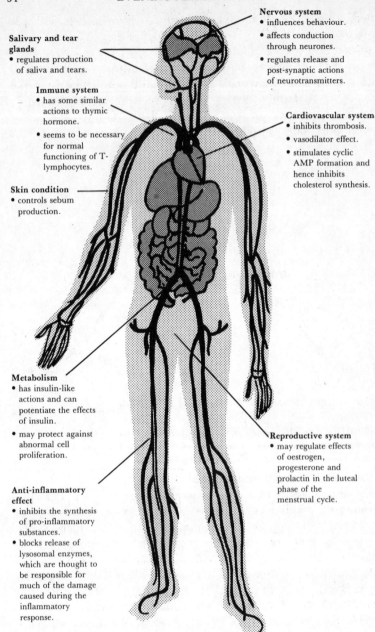

Nervous system
• influences behaviour.
• affects conduction through neurones.
• regulates release and post-synaptic actions of neurotransmitters.

Salivary and tear glands
• regulates production of saliva and tears.

Immune system
• has some similar actions to thymic hormone.
• seems to be necessary for normal functioning of T-lymphocytes.

Cardiovascular system
• inhibits thrombosis.
• vasodilator effect.
• stimulates cyclic AMP formation and hence inhibits cholesterol synthesis.

Skin condition
• controls sebum production.

Metabolism
• has insulin-like actions and can potentiate the effects of insulin.
• may protect against abnormal cell proliferation.

Reproductive system
• may regulate effects of oestrogen, progesterone and prolactin in the luteal phase of the menstrual cycle.

Anti-inflammatory effect
• inhibits the synthesis of pro-inflammatory substances.
• blocks release of lysosomal enzymes, which are thought to be responsible for much of the damage caused during the inflammatory response.

Figure 4. How Prostaglandin E1 Works in the Body

about. PGE2 is one of these, which is why he's been singled out here from all the other relatives. He makes various entrances throughout the book, always in unfavourable circumstances.

PGE1

Out of all the prostaglandins researched so far, PGE1 seems to have the most highly desirable qualities. These are just some of its wonder-workings:

- Dilation of blood vessels.[7]
- Lowering of arterial pressure.[7]
- Inhibition of thrombosis.
- Elevation of cycle adenosine monophosphate (AMP).
- Inhibition of cholesterol synthesis.
- Inhibition of inflammation and experimental arthritis.[8]
- Activation of defective T-lymphocytes.
- Prevention of liver damage and withdrawal symptoms in alcohol-addicted animals.[9]
- Inhibition of abnormal cell proliferation.[1,10,11]
- Inhibition of platelet aggregation.[12]

It's worth mentioning again that PGE1 is derived from gammalinolenic acid, which is the active ingredient of evening primrose oil. And that's why evening primrose oil has been tested in so many trials and for so many different medical conditions. Just looking at the above list, you can see why evening primrose oil is being used in research to do with heart disease and strokes; arthritis; various diseases of the immune system, and cancer.

Note: PGE1 does exist as a drug, manufactured by Upjohn Ltd. But it can only be used directly in very special circumstances because it is so short-lived and has to be injected into the veins.

5. PREMENSTRUAL SYNDROME

A few years ago, women used to call those awful days just before their period 'PMT' (Premenstrual Tension). Nowadays it's gone up in the world and the condition has taken on the rank of a syndrome. This is because there's more to PMS than tension alone. In any case, it's so common that most women know perfectly well what PMS is without having to be told. In fact as many as 40% of women aged between fifteen and fifty get PMS symptoms of varying degrees, and about 10% get them very badly indeed.

PMS can happen any time from two to fourteen days before a period, and the symptoms are likely to be both physical and psychological. The physical symptoms are fluid retention; weight gain; swollen ankles, legs and sometimes fingers; painful breasts; a feeling of bloatedness in the tummy; headaches; backache; a dull aching pain; skin problems like acne and blotchiness, and cravings for particular foods.

Alongside these bodily signs go the mental symptoms: depression; tension; irritability; lethargy; irrational weeping; sudden tantrums; illogical reactions; lack of concentration; loss of confidence and low self-esteem.

Some women are unlucky enough to have bouts of all of these. But most women who are truly suffering from PMS at the very least feel depressed, irritable and bloated.

It's not uncommon to meet women whose moods have become so black and thunderous at this time of the month that they've made life a misery for everyone around them as well as for themselves. Children and husbands often bear the brunt of it, so PMS affects many more people than the sufferer herself. PMS has, thank goodness, now been taken seriously by doctors, and women no longer have to be turned away from their GPs with dismissive statements about it being just a woman's complaint.

Since the publication of research on evening primrose oil and PMS, the oil is gaining ground as one of the treatments to try, particularly when the standard hormonal treatments haven't worked.

Evening primrose oil and the treatment of PMS

During 1981, a study was done at one of the major PMS clinics in the country, at St Thomas' Hospital in London. In this study, sixty-five women with bad PMS were treated with evening primrose oil *(Efamol)*. All of them had tried one or more other standard treatment, and all of them had failed.[9]

The results were good. 61% of the women experienced complete relief of their symptoms, and 23% had partial relief. The other 15% said there had been no change as a result of taking evening primrose oil. One particular symptom — breast discomfort — was helped considerably, with 72% saying this had improved. Other common symptoms to show improvement after taking evening primrose oil were mood changes, anxiety, irritability, headaches, and fluid retention.

Dr Michael Brush, the biochemist who did the study, liked the idea of using a natural product with a nutritional approach as many of the drugs for PMS can have undesirable side effects.

In almost all cases, the starting dose was two capsules twice daily after food. Although a few patients were given treatment all through their menstrual cycle, most of them started treatment three days before the symptoms were expected to arrive and carried on until the start of their period. In a few very severe cases, the dose was increased to three capsules twice a day. Some of the patients were given Vitamin B_6 (pyridoxine) at the same time.

Case Histories

Case 1.   A twenty-six–year-old domestic manager, who gave a clear history of PMS for ten to fourteen days before each period. This had gone on for the last six years. The main symptoms were severe breast discomfort, irritability, tearfulness, and poor co-ordination and concentration. Various hormone treatments had been tried without success. B_6, 75mg twice daily, had shown good results at first. The woman improved when she took *Efamol*, three capsules twice a day, plus six tablets of *Efavite* a day.

Case 2.   A forty-year-old school helper. She had attended the PMS clinic for four years. She gave a history of irritability, anxiety, depression, poor co-ordination, loss of libido, and moderate fluid retention for two weeks before each period. She had tried a hormone treatment, an anti-depressant, and B_6 without success. But when she took *Efamol*, two capsules twice a day, plus one *Efavite* a day, plus 100mg of B_6 a day, her symptoms disappeared.

Case 3.   A thirty-four–year-old school secretary. She had suffered from PMS since coming off the pill five years previously, and it had worsened since being sterilized three years later. She complained of severe breast pain, swelling of the stomach and face, irritability and depression. This happened for the last fourteen days of a regular twenty-eight to twenty-nine day cycle. Tranquillizers, diuretics, and B_6 had not helped. She was put on *Efamol*, four capsules a day, plus B_6 (80mg a day). This improved her mood, the swelling and the breast discomfort.

Case 4.   A thirty-four–year-old housewife. She had suffered from PMS since the birth of her first child eight years ago. The symptoms were lethargy and mood swings; irritability, depression, general bloating, very painful breasts and loss of co-ordination for fourteen days before her period. Various drugs had not worked. *Efamol,* four capsules a day, plus *Efavite*, four tablets a day, have given her complete relief of her symptoms.

Why evening primrose oil works for some cases of PMS

The precise mechanism isn't yet understood for certain as to why evening primrose oil helps PMS in some women. One possibility, now confirmed at St Thomas' Hospital, is that women with PMS are low in essential fatty acids. The relevance of this to PMS may be that a shortage of essential fatty acids can lead to an apparent excess of the female hormone prolactin.[8] Prolactin produces changes in mood and in fluid metabolism, similar to those found in PMS. Body tissues may be abnormally sensitive to normal levels of prolactin when they are low in essential fatty acids and Prostaglandin E1.

It is thought that Prostaglandin E1 (derived, of course, from evening primrose oil) can damp down these effects of prolactin. PGE1 also has complex interactions with steroids. The net effect of this is that PGE1 helps to smooth out the actions of the rapidly-changing hormone levels in the second half of a woman's menstrual cycle.

Vitamin B$_6$

One of the co-factors in the metabolic conversion process of essential fatty acids is Vitamin B$_6$. It has helped many women who suffer from PMS. Some of the women in the St Thomas' study took evening primrose oil and Vitamin B$_6$ together, as well as the other co-factors in *Efavite* (Vitamin C, Vitamin B$_3$, and zinc). Some women did very well taking evening primrose oil and B$_6$ together.*

Some women may have below average levels of B$_6$, and this is thought to have a harmful effect on the control centres for the menstrual cycle, the hypthalamus and pituitary glands. This vitamin is needed for the delicate inter-relationship between these two centres. Without it, they don't work properly.

The contraceptive pill, with its synthetic hormones, can lower some women's level of B$_6$. Hormones may produce a condition where far more B$_6$ is needed than usual — as high as 100mg up to 500mg a day. (*Efavite*, used in the St Thomas' study, has 25mg B$_6$ per capsule.)

Overall, it seems that Vitamin B$_6$ is helpful to many women with PMS, and also to women who get depressed when taking the contraceptive pill.[3]

Other studies with evening primrose oil and PMS

The PMS Clinic at St Thomas' Hospital is doing further

Efamol PMP: Britannia Health Products have put on the market a Pre-Menstrual Pack (PMP) especially designed for PMS sufferers. The pack contains 40 capsules of *Efamol* 500 (evening primrose oil 500mg and Vitamin E 10mg) and 20 tablets of *Efavite*, a mineral and vitamin supplement with B$_6$, Vitamin C, B$_3$ (niacin) and zinc as vital co-factors. The dose is two capsules of *Efamol* and one *Efavite*, twice daily for 10 days leading up to the period.

scientific trials at the time of writing. Women's Health Concern (address on page 93) is also involved in a large-scale study of *Efamol* for PMS.

In January 1982, the magazine *Here's Health* invited readers to take part in a two-month trial to test evening primrose oil for PMS. They used *Efamol*. The results of this survey were that 77% were less irritable, and 71% less depressed after taking evening primrose oil. 64% noted an improvement in breast pain and tenderness. Also, many women reported less pain and bleeding during their period.

Dr Caroline Shreeve, author of *The Premenstrual Syndrome — The Curse that Can be Cured* (Thorsons 1983) reports that eight out of ten of her patients with PMS improved as a result of taking *Efamol*.

Side effects. In the St Thomas' study, three patients complained of minor skin blemishes during treatment, and three had passing phases of excessive mood quietening. The other patients had no complaints of side effects. In the *Here's Health* survey, a few women reported side effects of headaches.

6. BENIGN BREAST DISEASE

One of the most successful trials on evening primrose oil so far published is for benign breast disease. In this disease, the breasts feel lumpy and tender, and they have a granular sort of texture. The breasts can be painful and swollen, with inflamed fibrous tissue and cysts. All these symptoms can be worse just before a period begins.

The research on evening primrose oil and painful breasts

A very thorough scientific trial at Ninewells Hospital, Dundee, together with the Welsh National School of Medicine took place in 1982.[10] Seventy-two patients took part; forty-one of them had symptoms which related to their periods (cyclical),* and thirty-one had symptoms which did not seem to be connected with periods (non-cyclical).

The results were very good, particularly for the women whose symptoms were cyclical. In this group, the tenderness and lumpiness of their breasts was significantly reduced after three months on *Efamol*. The women who were given a dummy oil (placebo) did not improve at all.

Women who had non-cyclical symptoms noticed a marked improvement in their breast tenderness after three months Efamol treatment, but their symptoms of lumpy breasts weren't helped during the period of treatment. However, the women on the placebo didn't see any improvements in either tenderness or lumpiness in their breasts.

The two surgeons who ran the trial, Mr Paul Preece in Dundee and Mr Robert Mansel in Cardiff, said '*Efamol* taken

---

*Painful breasts can be a symptom of PMS (see previous chapter). In the St Thomas' Hospital study, Dr Michael Brush found that 6 out of 9 cases of severe breast discomfort with fibrous cysts responded to evening primrose oil treatment.

with Vitamin C results in a significant reduction in both subjective symptoms and objective signs in patients with the cyclical pattern of benign breast disease.'

The beneficial effects of evening primrose oil don't happen suddenly. The symptoms get better gradually. The trial lasted for three months, but it's interesting to see that some women who stayed on the treatment for nine to twelve months got better and better, to the point where their symptoms vanished completely.

During the trial, the women were given two capsules of evening primrose oil 500mg (*Efamol*) three times a day. As the doctors found that women did better the longer they stayed on the treatment, they recommend that a woman tries the evening primrose oil for at least three to four months.

Side effects. Two of the women on evening primrose oil complained of gaining weight. One complained of a skin rash. On the other hand, several women volunteered that they felt generally fitter in themselves while on evening primrose oil. Moreover, some women said they'd built up a resistance to the common cold which they hadn't had before. The evening primrose oil had no toxic effects on anyone.

How evening primrose oil works in benign breast disease

There are still many uncertainties about this. One known fact is that a shortage of essential fatty acids in the diet does lead to excessive amounts of fibrous tissue.

Cysts, which are another common symptom of benign breast disease, may form because the body is for some reason making too much of the hormone prolactin and also is short of Prostaglandin E1. As a group, women with benign breast disease do seem to have higher levels of prolactin in their blood than they should.

The idea behind giving evening primrose oil as a treatment is that PGE1 can dampen down the effects of prolactin, may help prevent the development of cysts, and can help remove lumpiness in the breasts. The active ingredient of evening primrose oil is gammalinolenic acid, which is a precursor of Prostaglandin E1.

Coffee, tea and cola

The methylxanthine group of substances (caffeine, theophylline) found in coffee, tea and cola drinks increases the binding of prolactin to the breast. Removal of these stimulants from the diet has improved benign breast disease.

Conclusion

Paul Preece and Robert Mansel concluded the paper on their trial by saying:

> In practice, *Efamol* provides safe, simple, inexpensive therapy for a common problem. The modest price and its apparent lack of side-effects or toxicity make it very suitable for long term use. Its origin as a natural plant-derived substance (as opposed to a synthetic 'drug'), makes it highly acceptable to most patients. For women who have painful and lumpy breasts in a cyclical pattern over a protracted time period, and who, after appropriate investigation and reassurance still complain, *Efamol* is well worth trying in the first instance.

CAUTION: Any woman who finds a lump in her breast should consult her doctor first, and not treat herself.

Breast size

An interesting, unexpected and usually welcome effect of evening primrose oil in some women is that it seems to make their breasts bigger. No one knows exactly why. This was first reported in 1981 at a symposium on evening primrose oil when several women, quite unsolicited (and unsoliciting), revealed that they had gone up several bra sizes since first taking evening primrose oil — but without putting on weight anywhere else. The journalist Pearl Coleman was the first to put this beneficial side effect on the record, but her evidence was supported by several other women too. All the evidence so far has been purely anecdotal, and there have been no studies to investigate this phenomenon.

The two surgeons who conducted the benign breast disease trial, Paul Preece and Rober Mansel, remarked that they had seen no evidence of any increase in breast size in any of the women on their trial. Dr Mike Brush of St Thomas' Hospital, on the other hand, says there were some remarks about this happening among the women in his PMS group. However, it's worth noting that the women who have reported bigger breasts have usually been taking evening primrose oil for more than six months; in some cases for years.

Note: In the time I have been taking evening primrose oil (since 1974) I have gone up from a 36B to a 38DD without putting on weight anywhere else.

7. HEART DISEASE, VASCULAR DISORDERS AND HIGH BLOOD PRESSURE

Heart disease and the diseases of blood vessels are one of the biggest scourges of the Western world. If a way could be found to reduce the risk of getting them, then countless lives could be saved.

How evening primrose oil could help
The active ingredient in evening primrose oil is gammalinolenic acid, which converts inside the body to Prostaglandin E1. And three of the talents of PGE1 are that it:

* Lowers cholesterol levels
* Stops blood platelets clumping together
* Lowers high blood pressure

This has been seen in several clinical trials on polyunsaturates.[1,2,3,4,16] They all consistently show that polyunsaturates containing essential fatty acids, of which evening primrose oil is one, can lower cholesterol, reduce the clumpiness of platelets, and lower blood pressure.

Only diets which are rich in essential fatty acids and low in cholesterol, saturated fats and trans fatty acids (see page 27) are likely to be of any use whatsoever in reducing the risk of heart attacks and strokes. Evening primrose oil is a biologically active polyunsaturate.

Many oils on the market, especially processed vegetable oils and some margarines, are not polyunsaturates at all. Look carefully at the label. Once they have been processed, hydrogenated and deodorized they may lose their talents as essential fatty acids and actually turn into trans fatty acids, which behave as if they were saturated fats and act as blocking agents in the conversion process of essential fatty acids. They are unwise foods for people who risk getting any cardiovascular disease.

Lowering cholesterol

Evening primrose oil, of which the active ingredient is gammalinolenic acid, is better at reducing cholesterol in the blood than linoleic acid. This is what some research trials have shown so far: out of forty patients, the average fall in total cholesterol was 18% in men who took eight capsules of *Efamol* a day. This compared with 10% in men who took much higher doses of linoleic acid.[9,10] In one study it was calculated that GLA was about 163 times more potent than linoleic acid.[18]

It takes up to twelve weeks for evening primrose oil to reach a maximum effect in lowering cholesterol. Evening primrose oil has no cholesterol-lowering action in people with low cholesterol levels.

Blood clotting, thrombosis, and heart attacks

If you cut yourself and bleed, the blood soon clots and the bleeding stops. This is what should happen. But the blood shouldn't clot while it's travelling around inside you. When it does, there's trouble.

The clotting agents in the blood are called platelets. Platelets are tiny particles that look like plates. When you cut yourself they stick together so that a clot forms.

In healthy people, the blood only feels sticky when it begins to clot after a skin wound. If you could feel healthy blood while it was coursing round your body, it would be slippery. But in recent heart attack victims, the blood in their bodies has been found to be about $4\frac{1}{2}$ times stickier than in normal people. If you looked at their blood under a microscope you would be able to see platelets sticking to each other and to artery walls. When platelets stick to cholesterol deposits it quickly leads to a clot, which can block the blood flow. When a blood clot forms in an artery or a vein, it's called a thrombosis. This blocks the circulation in the area. A clot in a coronary artery is a coronary thrombosis. In the brain it's a stroke.

Normally, artery walls make prostacyclin which prevents platelets from sticking to each other or to artery walls. But arteries which are damaged by fat and cholesterol deposits, high blood pressure or injury, don't make enough prostacyclin.

How can evening primrose oil help? Diets rich in polyunsaturates reduce the clumping together of the platelets and lower the risk of thrombosis in both animals and humans.[4,13,14] In animals, evening primrose oil also increased the production of an anti-clumping factor by blood vessel walls.[15]

Gammalinolenic acid, which is the active ingredient of evening primrose oil, is a powerful anti blood clotter in surprisingly small amounts.[17] This is what Dr Alex Sim, head of Inveresk Research International, found when he was commissioned by Bio-Oil Research Ltd., to do some tests on evening primrose oil. As a result of his studies, Dr Sim believes that evening primrose oil could help people who are identified as being at risk of having a coronary thrombosis.

Experiments in test tubes, on baboons and in humans have all had the same results: the platelets stopped sticking together. And the un-clumping of platelets is one job which evening primrose oil does particularly rapidly.

Blood pressure

This is the pressure at which the heart pumps blood into the major arteries. If the blood pressure goes too high, it overtaxes the heart and blood vessels. People with high blood pressure run a greater risk of experiencing arteriosclerosis, heart failure, stroke, and kidney disease.

How could evening primrose oil help? Tests on both animals and humans have shown that essential fatty acids reduce arterial pressure.[11] Evening primrose oil has been shown to bring down the blood pressure in animals with high blood pressure.[12] In preliminary studies on humans with high blood pressure, evening primrose oil (given as *Efamol*) was considered more effective in lowering blood pressure than much higher doses of other polyunsaturates.

8. OBESITY

The discovery that evening primrose oil can help some obese people lose weight was found out quite by chance. During a trial on evening primrose oil for schizophrenia at Bootham Park Hospital in York it was discovered that several patients who were more than 10% above their ideal body weight lost weight while taking evening primrose oil. There had been no changes to their diet.[6] The evening primrose oil had no effect on people who were within 10% of their ideal body weight.

As a result of that chance finding, a group of investigators at the Metabolic Unit at the University of Wales in Cardiff set about investigating evening primrose as a treatment for metabolic causes of obesity.[9]

Brown fat (adipose tissue)

This is one of the key factors which explains why some people lose weight and others don't. The body has a special tissue known as brown fat which is found mainly in the back of the neck and along the backbone. The brown colour is due to the high concentration of cellular energy-producing (fat-burning) units called mitochondria. The brown fat burns calories not to produce energy for body movement, but solely for heat. One role of brown fat is weight stabilization; another is adaptation to cold weather. When brown fat is working normally it burns up any excess calories. But when brown fat is not working normally those calories are laid down as fat.[1,2] Some obese people have underactive brown fat.

Interestingly, the essential fatty acid content of body fat is inversely proportional to body weight.[5] This means that the higher the level of essential fatty acids in the body, the lower the body weight; and vice versa.[5]

How evening primrose oil helps brown fat work properly

The gammalinolenic acid in evening primrose oil has a stimulating effect on brown fat tissue. The prostaglandins which are the end-products of evening primrose oil metabolism possibly accelerate the mitochondria activity in the brown fat.

Sodium-potassium-ATPase

This complicated-sounding substance is an enzyme which is essential to the storage and transfer of energy in living cells and is found in all cell membranes. It is responsible for more than 20% of the total energy used by the body.[3,4] Its activity is low in many obese people. Studies have shown that there are decreased concentrations of sodium-potassium-ATPase in various tissues of obese mice. Along with defective brown fat, defective activity of sodium-potassium-ATPase is a major cause of obesity. Evening primrose oil activates this enzyme in cases where its activity is low.

The results of the Cardiff pilot study

The Cardiff doctors found that 30% of their patients had low sodium-potassium-ATPase activity, and they felt that this group had a metabolic cause for their obesity.[8] They investigated the effects of evening primrose oil (*Efamol*) in ten obese patients on a normal diet, against eight obese patients on a calorie-restricted diet. After six weeks, the Efamol group lost an average of around 5 kilograms, and their sodium-potassium-ATPase activity increased by an average of 52%. The group on th calorie-restricted diet lost an average of around $6\frac{1}{2}$ kilograms. Their ATPase fell by 37%.

The doctors concluded that through some unknown mechanism evening primrose oil might activate the mechanism involved in sodium-potassium-ATPase, so helping the body burn up excess calories.[7]

The most effective dose used was eight capsules a day. The people who are most likely to lose weight on evening primrose oil are those with low sodium-potassium-ATPase activity. This, and the activity of brown fat, can be tested by doctors in metabolic units of hospitals. A double-blind placebo-controlled trial of twenty-four obese patients is currently being conducted at the University Hospital of Wales.

9. ECZEMA, ASTHMA, ALLERGIES AND CYSTIC FIBROSIS

On the face of it, these all sound very different conditions. But in fact they have a lot in common — they are all to do with an abnormal body defence system. Doctors call this condition 'atopy'.

Most people call eczema plain simple 'eczema', but skin specialists divide eczema into two main types; contact eczema (or dermatitis) is quite simply an inflammation caused by contact with an irritant, such as a chemical or cosmetic. Atopic eczema is a chronic, patchy, mild inflammation of the surface of the skin which almost always begins in childhood. It can be made worse by irritants, but often occurs in the absence of any apparent cause.

The main symptom of atopic eczema is usually itching, often out of all proportion to the apparent severity of the rash. For most sufferers, atopic eczema is a mild affliction for most of the time, but it is liable to flare up on occasion. For some, however, it is a lifelong disease which blights their lives.

The condition behaves like a type of asthma where the patient is a little short-winded virtually all the time, and has an occasional attack of real difficulty in breathing. Atopic eczema is often associated with other more definite forms of allergy such as hay fever. So atopic eczema is part of a more general allergic response, which happens to show itself in the skin but might just as well take the form of , say, asthma.

In fact atopy — or a generalized allergic response — can show itself as any or all of a variety of conditions; eczema; asthma; hay fever; allergies (e.g. to house dust) and migraine among them. Atopy is common in patients with ulcerative colitis, Crohn's disease, ear problems, nasal polyps, and some obstetric problems. As many as one in five of the population suffers from some sort of atopy (though this term is virtually unknown by the layman).

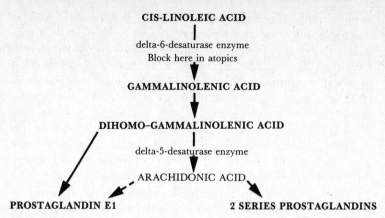

Figure 5. The importance of the Delta-6-desaturase Enzyme in the Conversion of Essential Fatty Acids.

People with eczema, asthma, allergies and atopic conditions have many things in common:

Faulty immune response. It has been known for a long time that people with eczema, asthma, and allergies have something wrong with their immune system — the defence system of the body. Allergic people form antibodies against, for example, articles of food, or clothing, or pollen. An antibody is a protein made in the blood-forming tissues in response to contamination with a substance known as an antigen. It activates the antigen by combining chemically with it, forming an inert compound.

Healthy people develop immunity to infectious diseases by forming antibodies that put the infecting microbes and their poison out of action. But allergic (atopic) people react to innocent intruders as healthy people react to dangerous invaders. Interestingly, atopic people are usually resistant to infections by intestinal worms and skin parasites. Atopy may be a relic of times when such defence was much more important than now.

Faulty enzyme function. Very recent research by Drs John Burton and Steven Wright in Bristol has revealed that atopic people may have a defect in an important enzyme, called the delta-6-desaturase.[4]

This is the enzyme which converts linoleic acid to

gammalinolenic acid. It has been found that patients with an atopic condition have normal or even high levels of linoleic acid, which shows they are eating enough foods with this essential fatty acid. But they are *very low* in gammalinolenic acid, and low in dihomo-gammalinolenic acid, and also in arachidonic acid.

This is explained by a block in the enzyme process for converting linoleic to gammalinolenic acid (see Figure 6). This block is probably the crucial defect in atopic conditions, such as eczema, asthma, and allergies. And it also helps to explain why the immune system is not working properly, as essential fatty acids are needed for it to function normally.

This is the only enzyme block which has been found in atopics. There is nothing wrong with the delta-5-desaturase enzyme, which helps convert DGLA to arachidonic acid.

How evening primrose oil helps

There was recently a trial involving fifty young adults with atopic eczema.[2] Most of them improved on evening primrose oil. At the beginning of the trial, all the patients gave blood samples. These were analysed, and the results showed that twenty-six out of the fifty had unusually high levels of linoleic acid. But gammalinolenic acid was found in only three of the fifty samples. Dihomo-gammalinolenic acid and arachidonic acid levels were also well below normal.

All the patients in the trial took evening primrose oil (*Efamol*) for twelve weeks. After that, their blood was analysed again. This time, the results showed a big rise in dihomo-gammalinolenic levels. There was also a steep rise in the level of arachidonic acid. At the end of the twelve week period, gammalinolenic was detectable in only eight of the fifty samples. But this deceptively low figure is only because GLA converts quickly and easily to DGLA. So the DGLA value is more important here. Even so, the levels of gammalinolenic and dihomo-gammalinolenic acid were still well below normal at the end of the twelve week trial. And the improvement was only seen in patients who had been taking a dose of eight or twelve capsules a day. Anything less than that did not work.

It seems from the results of the trial at Bristol Royal Infirmary that people would need to be taking eight to twelve capsules of evening primrose oil a day for a lot longer than twelve weeks in order to make up for decades with a faulty enzyme.

It seemed clear from the Bristol study that the culprit was the faulty enzyme. Evening primrose oil does nothing to correct the actual enzyme. But, by starting at the next step, it gives the body enough essential fatty acids, which can then turn into the valuable Prostaglandin E1.

Another factor that people with atopic conditions often have in common is food allergies. This is because their intestines are more permeable than other people's.[10] Essential fatty acids have a vital role in keeping surfaces impermeable.[11] It is known that when there is a shortage of EFAs, membranes become 'leaky'. This is another reason why evening primrose oil is so valuable in atopic conditions.

Evening primrose oil also helps correct the faulty immune system in people with atopic conditions. This is because evening primrose oil converts to prostaglandin E1, which stimulates the T-lymphocytes, which play a key role in the immune system. T-suppressor lymphocytes are a type of white blood cell which seems to keep the other parts of the immune sytem under control and which makes sure that the system first and foremost attacks foreign invaders, like bacteria and viruses, and not the body's own tissues.

It seems that the T-lymphocytes, especially T suppressor cells, are faulty in people with atopic conditions.[5,6] When T-suppressor cells are defective, auto-immune damage often happens.

Babies with eczema

Breast feeding is able to protect against the development of atopy.[12,13] Human breast milk, unlike cow's milk, is rich in gammalinolenic acid, dihomo-gammalinolenic acid, and arachidonic acid. A baby who suffers from an atopic condition probably can't make normal use of the linoleic acid found in cow's milk, because of the defect in the delta-6-desaturase enzyme. A baby who has this enzyme block may become allergic to certain foods, and develop allergic reactions to everyday things. It is also quite likely to develop eczema when it is switched from breast milk to cow's milk.

Babies and young children who suffer from atopy are more at

risk of getting viral infections, and they react badly to vaccinations, such as smallpox.

Evening primrose oil can safely be given to babies and young children, but it should be rubbed into the soft parts of the skin, like the inside of the thighs and on the tummy. The oil penetrates the skin very quickly. As a matter of interest, a 6-month-old, fully breast-fed baby is getting the equivalent amount of gammalinolenic acid as about 3 x 500mg capsules of evening primrose oil a day.

Cystic fibrosis

The findings of the Bristol doctors are also relevant to cystic fibrosis. They add weight to a suggestion made by two research groups, one in London and the other in Chicago, that children with cystic fibrosis have a severe block of the delta-6-desaturase enzyme — much more severe than in atopy — and this makes them extremely deficient both in essential fatty acids, and in Prostaglandin E1.

Children with cystic fibrosis also have a problem absorbing fats. They do not respond to treatment with safflower oil, which is rich in linoleic acid.[3]

Some people with an atopic condition are thought to be carriers of cystic fibrosis. An atopic parent could be said to have a single dose of abnormality in the enzyme block. A cystic fibrosis child has a double dose of abnormality.[1]

Delta-6-desaturase enzyme inhibitors

People with atopy or cystic fibrosis must be more careful about the things which can cause disruption to an already defective enzyme sytem.[17] These are:

- Trans fatty acids.[18]
- Catecholamines - hormones released by the adrenal glands during stress.
- Alpha-linolenic acid,[19] found in e.g. soy and linseed oils.
- Simple carbohydrates which produce a rapid rise in blood glucose.
- Alcohol.[20]

10. HYPERACTIVE CHILDREN

The signs of a hyperactive child are constant fidgeting and restlessness. His behaviour is abnormal and exasperating, and usually the cause of great distress to his parents. He is often troublesome and violent. A hyperactive child — even a toddler — can cause unbelievable havoc. At worst, he can wreck the house. At best, he's clumsy, uncoordinated, tearful and difficult. He seems to need little food and little sleep, and can drive everyone berserk when he starts tearing around the house at four o'clock in the morning.

At school, a hyperactive child is often singled out as 'disruptive'. Expulsions are not unusual. Hyperactive children can have high IQs, but often get poor results in their school work because they can't concentrate, and are slow in learning to speak.

Another tell-tale sign is abnormal thirst,[4] and hyperactive children are also more likely than other children to suffer from headaches, asthma, eczema and catarrh. In other words, hyperactive children are often atopic (see previous chapter). Most hyperactive children are boys. And an interesting observation made by the Hyperactive Children's Support Group (see page 93 for address) is that there are a lot of fair and ginger-haired children who are hyperactive. No one knows why.

It's often the mother who is blamed for the child's bad behaviour. She is accused of being at best a poor parent, and at worst a child batterer, or else of having done something to cause deep-seated psychological damage to the child.

This is rarely the case, although the constant broken nights and perpetual turbulence of the child would be enough to drive any mother to her wit's end. It's a great relief to mothers to find out that their child's hyperactivity isn't their fault at all, but may be due to a dietary and metabolic deficiency which can be corrected.

This dietary and metabolic deficiency may have a great deal to do with essential fatty acids. The Hyperactive Children's Support Group, having done a great deal of research amongst their own members, believe that two key problems with many hyperactive children may be a shortage of EFAs, and a sensitivity to certain foods and food additives. This shortage of EFAs was recently proven to be the case in a study at the University of Auckland.

There are several reasons for this. Firstly, it is known that hyperactivity is more common in boys than girls. And it's also known that males need about three times as much EFAs as females. Secondly, many of the foods and additives which are thought to lead to hyperactivity are the very things which are known to get in the way of the conversion process of essential fatty acids. Amongst them is tartrazine, an orange-coloured dye used in products like lollypops, and Ponceau 4R, another dye.

The food and food additives explanation for hyperactivity comes from American Dr Ben Feingold. The 'Feingold Diet'[1,2] recommends a diet free from food additives, certain preservatives, and foods containing natural salicylates found in fruits like apples, apricots, grapes, oranges, nectarines, peaches and plums.

Hyperactive children are sensitive to certain foods and food additives, and normal children are not. This biochemical difference between them can be explained by a shortage of EFAs in hyperactive children. So, the food and food additives theory fits in very well with the theory that hyperactive children are low in essential fatty acids: in addition to tartrazine, there are many substances eliminated by the Feingold Diet which work as weak inhibitors in the conversion of linoleic acid on its metabolic pathway to prostaglandins (see Figure 6). This means that prostaglandins are not being produced in the way they need to be for all the organs of the body to work properly. Prosaglandins seem to be particularly important in the brain, and have dramatic effects on behaviour.

It seems, however, that the foodstuffs banned in the Feingold Diet block PG formation only when EFA levels are low, and have little or no effect when EFA levels are high. This suggests that hyperactive children might have low levels of EFAs in their bodies. However, it is unlikely that the hyperactive child is not eating enough foods with EFAs in them. After all, he may well

CIS-LINOLEIC ACID

BLOCKING AGENTS:
  Trans fatty acids
  in processed foods.
  Saturated fats.
  Deficiencies of
  zinc, Vitamin B_6
  or magnesium

GAMMALINOLENIC ACID

DIHOMO-GAMMALINOLENIC ACID

BLOCKING AGENTS:
  Salicylates
  Tartrazine dye
  Ponceau R dye
  Opioids of Wheat & Milk*

PROSTAGLANDIN E1

Figure 6. Blocking Agents in Hyperactive Children

*Some hyperactive children are very sensitive to wheat and milk. The gluten found in wheat and the alpha casein found in milk can both give rise to opioid-like substances in the intestines called exorphins which can block PGE1 formation, especially if the EFA levels are low.

have brothers and sisters who are eating the same foods as he is, yet are perfectly normal. This points to a flaw in the hyperactive child's metabolic system.

There may be a variety of reasons why the hyperactive child is not absorbing EFAs normally. It may be partly genetic; it may be partly to do with the sex of the child (remember, males make less GLA than females); it may be that commercial foods made with saturated fats and processed vegetable oils are blocking the conversion of linoleic acid to GLA — that vital first step. It may also be that these children are lacking the essential co-factors needed for the metabolic conversion process. Studies by the Hyperactive Children's Support Group have shown that more than half their children are deficient in zinc.*

Another clue which leads to the theory that hyperactive children are low in EFAs is that many hyperactive children get very thirsty, and it is known that one of the first things that happens to animals on a diet short of EFAs is that they get thirsty.

It's also quite common for hyperactive children to suffer from complaints like eczema, asthma, allergies, and ear, nose and throat infections. It is known that EFAs are necessary for normal skin and normal body defences (see Chapter 9).

Once the amount of EFAs in the body is adequate, the child is no longe susceptible to the various food additives which block prostaglandin E1 formation. It has been found that children on evening primrose oil were able to go off the Feingold Diet and go back to eating ordinary food without any ill effects. However, the Hyperactive Children's Support Group recommend that children belonging to their members stay on the Feingold Diet and use the evening primrose oil as a safeguard for the times when the child breaks the diet, such as at parties, on outings, etc.

Studies on hyperactive children with evening primrose oil

Preliminary studies of the effect of evening primrose oil on hyperactive children have been carried out in the U.K., Canada, the U.S.A. and South Africa. All the doctors involved have reported some degree of improvement. Overall, about two-thirds of the children responded well. In some, the improvement was dramatic. In some children, asthma, allergies and eczema cleared up as well as the hyperactivity. But the regime did not help all the children — notably those with no history of eczema, asthma or allergies — and the regime is not being put forward as a cure for hyperactivity.

Case Histories

Case 1. A six-year old boy with severely disturbed sleep and

*The Hyperactive Children's Support Group is involved in controlled clinical trials using evening primrose oil and vitamins B$_3$, B$_6$, C, and zinc.

disruptive behaviour at home and school. He was threatened with expulsion, and his parents given two weeks to improve things. After advice from a friend, the boy was started on evening primrose oil. Three capsules were cut open and the oil rubbed into the skin morning and evening. The school soon reported a dramatic change in the child, and rang the parents to say so.

Case 2. A six-year-old boy who suffered disturbed sleep for years. He was always restless, and had frequent stomach upsets. His concentration and speech were poor and he was abnormally thirsty. When he was taken off wheat products, he became a changed child: his sleep, behaviour and speech became normal. He was then put on evening primrose oil (*Efamol*) and *Efavite*, three of each morning and evening, taken by mouth. After two weeks on this treatment, he was given some wheat foods and his behaviour and pulse rate remained normal.

Case 3. An eleven-year-old boy with eczema, disordered sleep and severely disruptive behaviour. He was expelled from school. Even after being put on the Feingold diet, he still had bouts of disruptive behaviour. He was given three capsules of evening primrose oil each morning and evening, by mouth. His eczema improved, and after eight weeks of treatment his hyperactivity had gone away.

Case 4. A twenty-year-old youth with a history of disturbed sleep and behaviour from birth. He developed eczema and asthma during childhood. He had very serious problems at school. At sixteen he went on a modified Feingold diet and progressively improved. He was given three capsules of *Efamol* and three tablets of *Efavite* morning and evening. His parents soon noted that their son became more relaxed and self-confident.

Case 5. A ten-year-old boy with poor attention span, who had been partially deaf from birth. He rocked himself to sleep, was impulsive, aggressive, and uncooperative. He improved a lot on the Feingold diet, but was still hyperactive, and deteriorated quickly when he ate the wrong foods. He was given three capsules of evening primrose oil morning and evening, plus one multivitamin and one *Efavite* tablet each evening. His behaviour and attitude improved dramatically, and he was promoted to the top stream in the school.

Dose

The Hyperactive Children's Support Group has found the following doses to be the most effective:

Age under 2 years – Please consult the HCSG (address on page 93) as it is necessary to work out the individual requirements carefully.

Age 2 to 5 years - 2 evening primrose oil capsules of 500mg per day, rubbed into the skin. Plus 2 tablets per day vitamin C 250mg, vitamin B_3 15mg, vitamin B_6 50mg, zinc sulphate 5mg.

Age 6 to 7 years - 3 evening primrose oil capsules of 500mg per day, either taken by mouth or rubbed into the skin. Plus 3 tablets per day of vitamin C 375mg, vitamin B_3 22.5mg, vitamin B_6 75mg, zinc sulphate 7.5mg.

Age 7 upwards - 4 evening primrose oil capsules of 500mg per day, either taken by mouth or rubbed into the skin. Plus 4 tablets vitamin C 500mg, vitamin B_3 30mg, vitamin B_6 100mg, zinc sulphate 10mg. Dose can gradually be increased to 6 evening primrose capsules.

For each age group, take the tablets twice a day, once in the morning and once in the evening with food. Many hyperactive children have a problem in absorbing nutrients from the gut. In these cases it is better to rub the evening primrose oil into the soft parts of the body, e.g. abdomen, tops of thighs, inside forearms. The oil is very fine and can be rubbed in until it is completely absorbed.

To remove the oil from the gelatin capsule, first leave the lid off the container so that the gelatin case softens. When you want to use the oil, pierce the capsule with a strong safety pin, and gently squeeze the oil out. It is possible to use the oil drop by drop by this method if necessary.*

*It is now possible to obtain pure evening primrose oil in a bottle which has a dropper attached to the cap.

Side effects

Nausea is sometimes felt when taking evening primrose oil, but this can be avoided by taking the capsules with food. This does not happen with the rubbing in method. Rashes may appear on very delicate skin when the oil is rubbed into the abdomen. This is not harmful, but the best way to avoid this is to rub the oil into the thighs or inside the forearms.

CAUTION: Do *not* give evening primrose oil if the child is epileptic.

11. SKIN, HAIR, EYES, MOUTH AND NAILS

Skin: eczema and acne

When rats were first starved of essential fatty acids in their diet, at the very beginning of research into essential fatty acids,[5] it was discovered that their skin showed the most obvious signs of deficiency[1]. Sebum (grease) production increased, sebaceous gland size increased, and a condition resembling eczema developed. These features have obvious similarities with acne and eczema in humans.[2,3]

Evening primrose oil has been successfully tested for eczema (see Chapter 9). As for acne, several people receiving evening primrose oil for other reasons have reported relief from acne. This may be because they had been short of essential fatty acids as increased sebum production is a possible sign of this. The possibility that evening primrose oil and the other nutrients which help PGE1 formation may be good for acne is now being tested in a placebo controlled trial in Bristol. While it is very unlikely that evening primrose oil will cure acne, it may be helpful in some people.

Hair: dandruff and hair loss

The same poor rats who suffered skin problems on a diet deficient in essential fatty acids also suffered a dandruff-like condition of their fur, and loss of their coat. This suggests that essential fatty acids are needed for healthy hair. As yet, there have been no trials on humans with hair problems using evening primrose oil.

Eyes and mouth

There are two conditions which make the eyes and mouth painfully dry. Sjogren's syndrome is a failure of the saliva and tear glands coupled with connective tissue disease, and is a complication of rheumatoid arthritis (see Chapter 12).

Sicca syndrome is a failure of the saliva and tear glands without connective tissue disease. Both can make the eyes horribly red and inflamed, and the lack of saliva can affect the teeth (saliva helps cleanse the teeth), and the gastro-intestinal system.

There is some evidence that patients with these conditions may respond to treatment with essential fatty acids, vitamin B_6, and vitamin C.[4]

Brittle nails
Quite by chance, when people with dry eyes and mouth were treated with evening primrose oil, it was found that their brittle nails dramatically improved at the same time. The patients volunteered this information unsolicited.[3] In fact people who are taking evening primrose oil for other conditions sometimes remark how good their nails have become since starting on the oil.

Results of a study for dry eyes and brittle nails
A small study conducted by Dr A. Campbell, consultant physician at Hairmyres Hospital, East Kilbride, and Dr G. MacEwan, consultant opthalmologist, Gartnavel General Hospital, Glasgow, came up with the following results, which were good for both dry eyes and brittle nails.
All the patients were treated with:

Efamol 500mg, 2 capsules, 3 times a day.
pyridoxine 25-50mg a day.
Vitamin C, 2-3 grams a day.

Patient 1. This fifty-five–year-old woman was first seen in December 1979 with a six month history of dry gritty eyes and of brittle nails which were prone to split when she manicured them. On the above regime her eye condition gradually improved and after one month she spontaneously reported that her nails had become perfectly normal.

Patient 2. This seventy-two–year-old woman was started on treatment in November 1979 following an eight month history of dry, gritty eyes and a six month history of nails splitting and breaking off. Within one month her eyes improved and by February 1980 they were moist and had normal tear secretion.

At the same time her nails also became normal.

Patient 3.   This fifty-three–year-old woman had a history going back several years of defective tear secretion, and when she was seen in August 1980 she stated that her nails had been brittle and splitting for about six months. After one month of the above treatment her eyes had improved and her nails had stopped splitting.

Patient 4.   This forty-eight–year-old woman was diagnosed as having Sjogren's syndrome and rheumatoid arthritis. By December 1979, although her joint symptoms were variable and not particularly severe, her dry eyes and dry mouth had become marked and very troublesome. She had very brittle nails which could not be manicured and her hands and nails were very sensitive to exposure to detergents.

Therapy with non-steroidal anti-inflammatory aspirin-like drugs was stopped and she was started on the above regime. Over the next two months both her tear and saliva production improved and her nails, for the first time in many years, became hard and could be manicured normally. At the same time the sensitivity to detergents disappeared. This patient also had mild Raynaud's syndrome which did not appear to be substantially benefited by the treatment. In July 1980 therapy was stopped with the patient's consent. Her mouth became dry within one week and within two weeks she noticed that the skin around her nails had become dry and that her nails were beginning to split. Tear production seemed less affected. Treatment started again in September 1980 and again there was a rapid improvement in both her nail quality and the dry mouth.

In the light of these, and other case histories Drs Campbell and MacEwan suggest that the nutritional approach to dry eyes and brittle nails, using evening primrose oil, may be of value. Dr Campbell also thinks that poor nail quality is a new sign of EFA deficiency in humans. A fuller trial of evening primrose oil on Sjogren's syndrome and Sicca syndrome is underway at the Scottish centres. A placebo controlled trial, recently completed at the University of Copenhagen, has confirmed that *Efamol* is effective in increasing tear production.

Why evening primrose oil was used

It is known that when there is a real shortage of essential fatty acids, for whatever reason, it can lead to the failure of the immune system, and susceptibility to infection and inflammation.

In conditions like Sjogren's syndrome and Sicca syndrome there seems to be too much of the 2 series prostaglandins produced and not enough of the 1 series prostaglandins. But experimental studies have shown that disorders such as these may be corrected by Prostaglandin E1. The active ingredient of evening primrose oil is gammalinolenic acid, which converts into Prostaglandin E1.

Evening primrose oil in cosmetics

The Western world now seems to be catching up with what the American Indians knew centuries ago — evening primrose oil is good for your skin. Evening primrose oil is sold to some very classy cosmetics firms, who use it as an active ingredient in their face creams.

Barbara Cartland, the tireless authoress, promotes the F-500 cream and capsules for the Cantassium Company. She claims that using the cream and tablets combined 'recreates the ageing cells and I am absolutely convinced gives one "New Youth".'

Evening primrose oil has even got into soap. 'Satin Soap' with evening primrose oil and Vitamin E, manufactured by the Vale of Health Organic Products Co., has this to say for itself in its promotional literature: 'When applied to the face, the Vitamins E and FF in evening primrose oil helps to clear skin blemishes and wrinkles due to dryness. This natural oil regenerates cell growth, helps to stimulate the circulation of the skin and revitalises the ageing cells. It is used as a moisturiser and emollient. It helps maintain natural softness and suppleness of youthful skin.'

If you take the oil out of a capsule by puncturing the gelatin, you can apply it directly to your skin, and use it like a face cream.*

*Note: Evening primrose oil is now sold in a bottle complete with its own dropper.

12. RHEUMATOID ARTHRITIS AND OTHER INFLAMMATORY DISORDERS

Rheumatoid arthritis is a chronic disease affecting connective tissue, mainly of joints. It can be very painful, and affects about one person in twenty in Britain at some time. The figure rises dramatically after the age of sixty-five. Some people are seriously crippled by the disease.

Rheumatoid arthritis begins as an inflammation in the membrane — known as the synovial membrane — that produces the fluid that lubricates bone joints. The cause of the original inflammation is uncertain, but RA is classified as an auto-immune disease; one where the body is mistakenly attacking itself.

One characteristic feature is a small nodule of inflamed fibrous tissue just under the skin, which feels very tender. Any organ may be affected, and RA is a systemic disease, which means it can affect the whole body. However, the most common symptoms arise from inflammation of the fibrous connective tissue around the joints. The knuckle joints and wrists tend to suffer most. The disease is more common in women than in men.

The immediate damage to joints is thought to be due to the release of destructive enzymes, called lysosomal enzymes, from inflamed cells. It is these enzymes which are thought to be destroying cartilage and bone, though there is still research going on in this area.

A shortage of essential fatty acids, and inflammation

If your body is short of essential fatty acids it can lay you open to inflammation or infection, and your immune system is quite likely to go wonky too. That's how important essential fatty acids are!

In rheumatoid arthritis, a disorder of inflammation and immunity, it seems that the body is producing too many of the wrong kind of prostaglandins, and not nearly enough of the right ones. So there are 2 series prostaglandins in profusion, but not enough of the heroic PGE1.[1] And that's not surprising, because if there are no essential fatty acids getting through the system, no PGE1 can be made, as PGE1 comes from essential fatty acids and can't be made without them.

The hordes of 2 series prostaglandins are usually on the scene of crime when there's some local inflammation around. Unlike their good cousins, PGE1, there's no shortage of 2 series PGs because they're made differently. They're derived from arachidonic acid (found in foods like meat and dairy produce), whereas PGE1 owes its existence to the essential fatty acids which will metabolize into gammalinolenic acid and dihomo-gammalinolenic acid as long as they don't get lost on the way.

When PGE1 is in short supply, the immune system is one of the first to feel the pinch. After all, PGE1 has a very important job in controlling T-lymphocyte production,[9] and the T-lymphocytes are the body's crack troops. It's as if the front line has gone absent without leave, and the body gets muddled as to who's on it's side, and who it should be sending salvos to. So it looks as though the missing PGE1 has a lot to answer for: hordes of 2 series PGs, and the odd behaviour of T-suppressor cells. If only there were more Prostaglandin E1s around, none of this might have happened.

Unfortunately, there's no clear explanation as to why the PGs are out of balance in rheumatoid arthritis. It could be because not enough foods with cis-linoleic acid are being eaten. Or perhaps the crucial delta-6-desaturase enzyme is not working properly because of some blocking agents getting in its way. Or maybe there's a shortage of any or all of the important co-factors, Vitamin C, B_6, B_3, zinc, and magnesium.

Whatever the reason, it is possible to start up the manufacture of PGE1 once again, simply by giving evening primrose oil, which is a precursor of Prostaglandin E1.

What evening primrose oil can do

Evening primrose oil converts into Prostaglandin E1, which has the power to sort things out. Once PGE1 is on the production line again, the normal balance between the 1 and 2 series PGs can be happily restored. This is because when PGE1 levels increase, the 2 series automatically come down, like a see-saw.

PGE1 blocks the mobilization of arachidonic acid, and so reduces the level of 2 series PGs. So the hordes simply subside. And by making more PGE1, the lymphocytes are stimulated to start defending the body as they were trained to do.

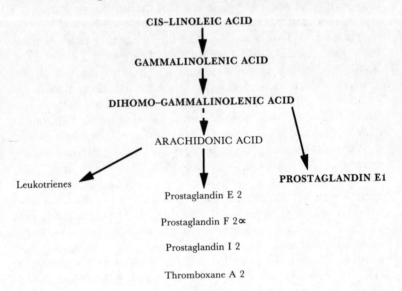

CIS–LINOLEIC ACID

GAMMALINOLENIC ACID

DIHOMO–GAMMALINOLENIC ACID

ARACHIDONIC ACID

Leukotrienes   **PROSTAGLANDIN E1**

Prostaglandin E 2

Prostaglandin F 2α

Prostaglandin I 2

Thromboxane A 2

Figure 7. How the 2 Series Prostaglandins and Leukotrienes are Made.

— —This step does not seem to be very active in adult humans. Most humans get most of their arachidonic acid directly in the diet from meat and dairy products. A diet low in meat and dairy products, and supplemented with evening primrose oil, may prove the best treatment for arthritis.

There are several ways in which evening primrose oil might work, if you want to examine its actions under the microscope. Some of these can only be described in rather technical language.

1. The human thymus and lymphocytes are relatively rich in PGE1.[2] Prostaglandin E1 has several actions which are similar to those of thymic hormone and seems to be necessary for the normal working of T-lymphocytes.[3,4] T-lymphocytes are known to be abnormal in rheumatoid arthritis and other inflammatory disorders.

2. Lysosomes are organelles in the cells which contain a variety of destructive enzymes. These enzymes can be released during inflammation and may be the cause of much of the damage. PGE1 can block the release of lysosomal enzymes.

3. Arachidonic acid is released from its cellular stores in most types of inflammation. It is converted to 2 series PGs and to substances known as leukotrienes, both of which are powerful inflammatory agents. Evening primrose oil, by converting to PGE1, reduces the conversion of arachidonic acid to 2 series PGs and to leukotrienes.

Research with evening primrose oil on inflammatory disorders

There have been several animal studies which raise the possibility that evening primrose oil may be helpful in human rheumatoid arthritis. Dr Robert Zurier and his group from Pennsylvania pioneered the idea that Prostaglandin E1 is anti-inflammatory. He and his researchers showed that PGE1 can inhibit experimental ('adjuvant') arthritis in rats; control the systemic-lupus-like syndrome in NZB/W mice; activate T-lymphocytes; and control lysosomal enzyme release in humans.[5,6] Zurier's group have now shown that evening primrose oil (*Efamol*) was as effective as PGE1 in controlling experimental arthritis in rats.[7] (In the first experiment PGE1 itself was injected into a vein. In the second experiment evening primrose oil was used, which converted into PGE1 inside the rat's body.)

Evening primrose oil has been found to restore tear and saliva production in some patients with Sjogren's syndrome, a type of rheumatoid arthritis. It also healed ulcers and relieved peripheral vascular spasm in some patients with systemic sclerosis, another disease related to rheumatoid arthritis.

Raynaud's syndrome has also been shown to respond to infusions of PGE1. Raynaud's syndrome is a disorder of small arteries, especially in the hands and feet. It is common in arthritis and related conditions. The arteries have a tendency to go into spasm, so the patient can suffer sudden attacks of pallor and extreme coldness in the extremities. The attacks are due to a reduction in the blood flow caused by spasm in the arteries.

The effects on patients with inflammatory conditions

The effects of evening primrose oil do not usually become apparent in less than four to twelve weeks. Paradoxically, some of the patients who turn out to respond to evening primrose oil best, actually get worse in the first couple of weeks. But this phase soon passes. There is some preliminary evidence that the course of the disease may actually be stopped in some patients, *although there can be no reversal of existing structural damage.*

A note about aspirins and other anti-inflammatory drugs. Drugs like aspirin and indomethacin, known as non-steroidal anti-inflammatory drugs, block the conversion of arachidonic acid to prostaglandins. And the steroid drugs block the release of arachidonic acid. This is how all these drugs have an anti-inflammatory action. But, unfortunately both sets of drugs also block the release of dihomo-gammalinolenic acid and its conversion to PGE1. So, while they do block the release of the 2 series PGs, they also reduce the amount of PGE1 being formed, and this process may be defective already. Gammalinolenic acid, the active ingredient of evening primrose oil, does not work in patients who are taking more than very small doses of these drugs. However, evening primrose oil can be given with the drugs for six weeks or so to build up stores of GLA and DGLA, and then the drugs can be cautiously tapered off.

Important Note:
It must be stressed that at the moment the use of GLA in arthritis is mainly based on animal studies. Careful placebo-controlled trials in humans are underway, but until they are completed nothing definite can be said about evening primrose oil and human rheumatoid arthritis.

13.   MULTIPLE SCLEROSIS

Multiple sclerosis is more of a problem than some other diseases because no one's sure what causes it, and because the disease can be so variable not just between one patient and another, but within the same patient at different times.

At its very worst, people with it can go completely gaga, unable to move, speak clearly, or do anything for themselves. At its best, people with MS can run around and lead perfectly normal lives. But most people who have MS have some trouble with walking, and their hands, eyesight and speech are often affected too, though their symptoms often come and go.

It is now thought that many factors play a part in MS. Many neurologists are convinced that a virus lurks somewhere, if only they could find it. Recently, other doctors have come up with clues which make them think that MS is a disease of the vascular system, and that the neurological damage is due to blood leaking into the brain through permeable blood vessel walls, triggered by some kind of blockage, possibly caused by fat. Blood in brain tissue is toxic, and can cause demyelination, which is one of the features of MS.

Other researchers have found evidence of a faulty fats metabolism in MS. As well as possibly being a metabolic disorder, MS is also classified as an auto-immune disease, where the body goes to war against itself. There is some evidence in MS that the brain is being attacked by certain white blood cells.

Multiple sclerosis is called a 'multi-factorial' disease, and it is possible, though not yet proved, that all the things mentioned above are happening. For example, if there is something wrong with the way the body is metabolizing fats, then this could lead to a fault in the immune system. This would allow viruses to stake a claim in the central nervous sytem, and could provoke an auto-immune attack. A defect in handling fats would also explain the leakiness of blood vessels walls and the damage to the myelin sheath.

Multiple sclerosis and fats

There are many facts which point to the theory that there could be something wrong with the way people with MS metabolize fats. It is known that the nerve tissues, plasma, and red cells of MS patients are low in linoleic acid and arachidonic acid.[15] Tests have shown that the levels of linoleic acid in the plasma tend to be low in some — but not all — MS patients, and they go even lower during an exacerbation.[1] (Low linoleic acid levels can also be found in several other diseases.)

The first big trial to do with linoleic acid and MS was done in 1973 by Dr J.H.D.Millar of Belfast and Dr K.J. Zilkha of the National Hospital in London.[2] They found that when linoleic acid, in the form of sunflower seed oil, was given to patients with MS it reduced the frequency and severity of relapses. This has been confirmed by another study.[3]

At that time, sunflower seed oil became all the rage with MS patients. And from this study, John Williams, the man who started Bio-Oil Research Ltd., had the idea that if sunflower seed oil helped MS a little, then evening primrose oil, being that much more biologically active, might help even more. It was a brainwave of an idea.

At around the same time Professor E.J. Field in Newcastle was doing some tests to find out the effect of linoleic acid on the lymphocytes (white blood cells) in MS. He was then in charge of the MS research at the Medical Research Council's Demyelinating Diseases Unit.

In MS, there is some evidence that one type of lymphocyte is actually attacking the brain. This is part of the auto-immune reaction in MS. In Professor Field's tests, it was found that linoleic acid had a dramatic effect in dampening down the lymphocytes;[16] it was thought that this stopped them attacking the myelin sheath in the brain. (It is the loss of myelin in the brain which causes the symptoms in MS. Like an electric cable where the insulation has been shaved off, impulses can't travel properly.)

Professor Field soon began tests with people who had MS. He took samples of their blood, and was looking out for the effects of evening primrose oil (taken as *Naudicelle* capsules) on their red blood cells. The results of these blood tests proved that gammalinolenic acid was much better than linoleic acid in correcting the defects found in the blood of MS patients.[17] Professor Field strongly recommended that people with MS should take evening primrose oil capsules. Ever since then, evening primrose oil has been widely used by MS sufferers in Britain. In fact, so far, evening primrose oil has been more widely used for multiple sclerosis than for any other condition.

Animal fats and MS

MS is much more common in countries where a lot of animal fat, or saturated fat, is eaten. And the disease is far less common in countries where they eat a lot of polyunsaturates. This might be purely coincidental, or it might mean that, for some reason, saturated fats are interfering with the normal metabolic process in people who get MS.

The geographical distribution of MS is the basis for the low saturated fat/high polyunsaturated fat diet recommended more than 30 years ago by Professor Roy Swank of Portland, Oregon. People with MS who have faithfully stuck to his saturated-fat-free diet have deteriorated far less than might otherwise have been expected.[4,5]

How evening primrose oil might be working in MS

1. It stimulates the T-lymphocytes. Prostaglandin E1 stimulates the normal function of T-suppressor lymphocytes.[11] These are cells which keep the other parts of the immune system under control and which make sure that the body's defences attack foreign materials and not the body's own tissues. When T-suppressor cells are defective, auto-immune damage frequently occurs. Research has shown that T-suppressor cells are very low in MS patients during a relapse, and Prostaglandin E1 may help prevent this. It is known that PGE1 has the effect of dampening down the B-lymphocytes which are capable of attacking the central nervous system.

2. It stops the platelets clumping together. In MS there is evidence to show that the platelets clump together in an abnormal way. The platelets are the small plate-like particles in the blood which help the blood clot. PGE1 regulates the platelets and stops them bunching up together, sticking to each other and to blood vessel walls.

3. *It makes faulty red blood cells return to normal.* In MS, red blood cells are not only very low in essential fatty acids, they are also much bigger than they ought to be, and have a poor ability to regulate the passage of fluids through cell membranes.

Dr Michael Crawford and Dr Ahmed Hassam showed that evening primrose oil can make red blood cells normal within a matter of months. Drs Crawford and Hassam have also done tests to show that essential fatty acids are necessary for brain growth and maintenance. As well as linoleic acid, they also believe that alpha-linolenic acid and fatty acids formed from it (found in foods like oily fish, fish oils and dark green leafy vegetables) are important in MS too.[23]

Evening primrose oil has also been shown to correct the defect in the mobility of the red blood cells, which Professor Field invented as a diagnostic test for MS.[6,21] This test, in a modified form, and known as the electrophoretic mobility test, has been repeated by other researchers[8,9] and most recently at Charing Cross Hospital in London.

In a recent follow-up study on MS patients on long-term treatment with evening primrose oil (*Naudicelle*), who were also following a diet low in saturated fat, it was found that the mobility of the red cells was restored to normal.[8] The most responsive cases were those who had experienced frequent relapses.[19]

4. *It strengthens blood vessel walls.* Prostaglandin E1 is known to strengthen blood vessel walls. This may be a very helpful thing to do in MS because there is some reason to believe that people with MS have such leaky blood vessel walls that they may be allowing blood to seep through them into the brain. Having stronger blood vessel walls also makes them more resistant to the bits and pieces (platelets, cholesterol) clumping together and sticking to them.[13]

5. *It possibly acts as an anti-viral agent.* When human cells become transformed by viruses, they always lose the ability to convert linoleic acid to gammalinolenic acid, and so can't make Prostaglandin E1.[12] This may make the transformed cells resistant to attack by the body's natural defences, the immune system. Evening primrose oil gets over this problem by its

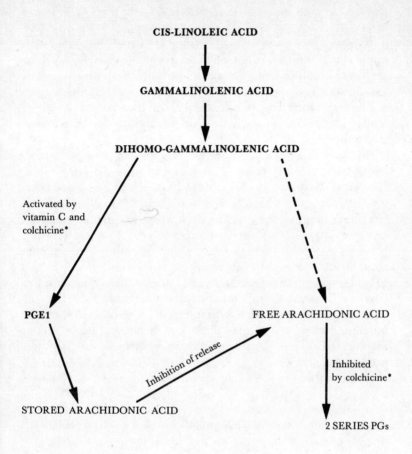

Figure 8. How evening primrose oil favours the PGE1 route.

*The drug colchicine is only available on prescription. It helps essential fatty acids select the pathway to PGE1 and away from the 2 series PGs.

ability to convert easily into Prostaglandin E1, and PGE1 may restore the cells' normal susceptibility to the body's immune sytem.

6. *It affects the nervous system.* Prostaglandin E1 has effects on nerve conduction and on the action of nerves. This can produce profound changes in the workings of both the central nervous system and the peripheral nervous system. PGE1 has strong regulating effects on the release of neurotransmitters at nerve endings and also on the post-synaptic actions of the released transmitters.

7. *It maintains a healthy balance between the 1 and 2 series prostaglandins.* If the body is very low in essential fatty acids, there is a sharp rise in the 2 series PGs, which are made from arachidonic acid. A high level of the 2 series PGs is a feature of various inflammatory disorders, such as rheumatoid arthritis (see previous chapter) and multiple sclerosis. It has recently been shown that the cerebrospinal fluid from MS patients contains high levels of PGF2alpha.[14]

When you increase the amount of essential fatty acids in the diet, this puts dihomo-gammalinolenic acid and PGE1 back in business. And once PGE1 is on the scene, it puts arachidonic in order and makes sure that there is a healthy balance in the amount of 1 and 2 series prostaglandins coming off the production line.

Another thing that evening primrose oil does is that it makes it more likely that PGE1 will be produced, as against PG2. At the junction where the road forks at dihomo-gammalinolenic acid, it persuades the traveller to follow the route towards PGE1 instead of taking the route towards arachidonic acid and the 2 series PGs.

Reported benefits of evening primrose oil on MS

During 1979, a survey was carried out by Bio Oil Research Ltd. to assess the views of MS patients taking *Naudicelle* as a dietary supplement.[20] 480 MS sufferers took part in the survey. Of these, 65% felt there was some improvement in their condition. Of these, 43% said there had been a stabilization of their condition - they had got no better, but they had got no worse. 22% said there had been fewer and less severe attacks. 20% said certain symptoms had been alleviated. 13% reported an improvement in general health. 2% reported further beneficial side-effects. The overall results look like this:

Some improvement —   65%
No change —   22%
Deteriorated —   10%
Don't Know —   3%

Improvements. People in the 'some improvement' category mentioned the following benefits:

- Increased mobility.
- Increased walking ability.
- Reduced spasm or tremor.
- Improved bladder function.
- Improved eyesight.
- Improved condition of hair and skin.
- Relief of constipation.
- Improvement in wound healing.
- Regaining correct weight.
- Heavy periods returned to normal.

Note: The 'improved' group contained a significantly higher proportion of MS patients who had been diagnosed within the preceding four years.

The ARMS Survey

Action for Research into Multiple Sclerosis is a self-help pressure group in Britain whose members all have MS themselves, or are related to someone with it. In 1977 ARMS sent out a questionnaire to all its members to find out what effect *Naudicelle* was having on them. They were also asked to get an opinion from their own GP as to their condition since taking the capsules. 177 completed questionnaires were returned. These were the results:

Improved —   127
No change —   33
Worse —   17

Of the 127 in the 'improved' category, there were fifty-nine testimonials from GPs supporting this assessment. (Not everybody who filled in the questionnaire bothered to see their doctor.)

Even though this survey has no scientific standing, and all the

answers are based only on the subjective opinion of the MS
sufferer who filled in the questionnaire, the results are
nevertheless extremely encouraging.

ARMS members were also asked how long they had been
taking *Naudicelle*. The answers showed that improvements
increased when they had been taking the capsules for more than
four months. Beneficial effects appeared as follows:

Under 4 months —   35%
4 months to 1 year —   73%
1 to 2 years —   73%
2 to 3 years —   82%

At the time of the survey, very few members had been on
Naudicelle for longer than three years.

141 of the people who returned the completed questionnaires
were also on some kind of diet. The results showed that the
people who were exercising some control over their diet (i.e.,
less saturated fat) had better results with the evening primrose
oil.

Clinical trials on evening primrose oil and MS

Two clinical trials by Drs D. Bates, P.R.W. Fawcett, D.A.
Shaw, and Dr Weightman in Newcastle in 1978 did not show
encouraging results. However, when these trials were
conducted the capsules contained two dyes, tartrazine and
ponceau R.,[3,10] which are now known to interfere with fatty acid
metabolism. (The oil now comes in dye-free clear gelatin
capsules, but no clinical trials have been done since this
improvement). Professor E.J. Field has now reported that when
dye-free capsules are used, reversal of red cell electrophoresis to
normal takes place within six to twelve weeks.

Note: All these studies took place before *Efamol* was available on the market.

Conclusion

Recent studies in America have shown that the people who
were most likely to benefit from essential fatty acid therapy were
those who were found to be deficient in them.[22] This was
discovered by doing a blood profile of the essential fatty acids.
Not everybody with MS had an abnormal EFA blood profile.

The other recent observation made in America is that the
people who benefit most from essential fatty acid therapy are

the newly-diagnosed, still in the early stages of the disease. Evening primrose oil can do nothing to reverse existing damage, but may help stabilize a condition that has not yet become bad. This is one reason why it is so important for doctors to tell their patients the diagnosis of MS, rather than keep it from them, and for nutritional treatment to start early.

Note: There is no exact animal model for MS. The nearest is called experimental Auto-Immune Encephalomyelitis (EAE). In controlled studies, evening primrose oil has been shown to inhibit the formation of this disease.[7]

Dose

Breakfast:	2 capsules evening primrose oil 500mg; 250mg Vitamin C; 50mg pyridoxine; 10mg zinc sulphate; 0.5mg colchicine.
Lunch:	2 capsules evening primrose oil 500mg; 250mg vitamin C; 50mg pyridoxine.
Supper:	2 capsules evening primrose oil 500mg; 250mg vitamin C; 50mg pyridoxine; 10mg zinc sulphate; 0.5mg colchicine.
Bed-time:	2 capsules evening primrose oil 500mg; 250mg vitamin C.

This is eight capsules of evening primrose oil of 500mg each per day. However, it is safe to take up to twelve or sixteen capsules if there is only partial improvement on the 8-a-day dose. Up to twenty capsules have been taken per day by volunteers without any apparent ill effects.

Evening primrose oil may do some good in MS, and cannot do any harm. It probably cannot help everyone with MS, but is certainly worth a try judging from the very beneficial effects noticed in many MS sufferers who have been taking the oil for years.

Patients should eat a diet low in animal and saturated fats to make this therapy more effective.

Note: I have been taking evening primrose oil for MS since 1974. My blood was analyzed then and found to be low in some of the essential fatty acids. When I started taking EPO my red blood cells gave an MS reading according to Prof. Field's test. Now the EFA profile of my blood is normal, and so are the red cells. MS symptoms are minimal.

14. SCHIZOPHRENIA

Schizophrenia is the most serious of all mental illnesses. It seems to be a biochemical disorder in which thought processes become very severely disordered.

Scientists aren't quite sure yet what the cause of schizophrenia really is, and a variety of suggestions have been put forward. Excess dopamine activity is one possible cause (dopamine is a compound involved in nerve transmission); another one is that schizophrenics are producing an abnormal opioid, or else producing too much of a normal opioid. Another suggestion is that schizophrenics are very low in Prostaglandin E1. Another is that they are hyper-sensitive to wheat proteins. It might be an allergic phenomenon, suggest others, or a defect in zinc metabolism, or a pineal deficiency.

Or perhaps it's all these things. Dr David Horrobin, writing in *The Lancet*, suggested that all these concepts could well be aspects of the same problem. But he put forward his own idea that the common path in schizophrenia may be the failure of schizophrenics to make PGE1, and the lack of PGE1 activity in their bodies.

Schizophrenia and prostaglandins

There is evidence that PGE1 levels may be low in schizophrenia. On the other hand, the levels of PG2 are high. In fact the cerebrospinal fluid from schizophrenics has been found to contain high amounts of it.[3,4]

Blood platelets in schizophrenics don't make PGE1 normally,[1] and red blood cells from schizophrenics have got vitually no DGLA and very little linoleic acid.[2] But they do contain three times the normal amounts of arachidonic acid.

Prostaglandin E1 is able to inhibit some effects of dopamine. The abnormal amounts of dopamine found in schizophrenics could therefore be something to do with them having too little

PGE1 to keep it in check.[5] The excess of dopamine is currently the most popular idea as to the mechanism of schizophrenia.

Evening primrose oil and penicillin in the treatment of schizophrenia

Evening primrose oil has been tried in some studies with schizophrenia because of its ability to convert to Prostaglandin E1. This prostaglandin is low in schizophrenics.

Evening primrose oil and penicillin had a modest therapeutic effect in six severely ill patients who were taking part in a preliminary study at Bootham Park Hospital in York, run by Dr K.S. Vaddadi. Previous studies with evening primrose oil in schizophrenia had suggested that, like penicillin, it could prevent a relapse, but could not bring about an improvement.* So they tried evening primrose oil and penicillin together.

As a result of the study in York, two schizophrenic patients who had been severely ill for twelve and twenty-one years respectively, in spite of different phenothiazine drug regimes showed a really dramatic improvement after treatment with evening primrose oil and penicillin together. No patients got worse during the sixteen weeks of the study.**

Dr Vaddadi thought his results were certainly good enough to do further research, and a bigger controlled trial is currently underway.

Case studies

Case 1. A divorced woman aged twenty-one with two children. She was admitted to hospital in 1957 with a diagnosis of schizophrenia. She had neglected to take care of either herself or her home. She was withdrawn, and made only monosyllabic replies. She had paranoid and hypochondriac delusions. Treatment with drugs was only moderately successful. In 1968 she became catatonic, and was treated with ECT (electro-convulsive therapy). This gave a poor response, and over the

---

* A placebo-controlled trial of penicillin in schizophrenia showed that it was modestly effective.[6] Evening primrose oil, either alone or in combination with penicillin or the co-factors was effective in treating some chronic schizophrenics.[7,8]

**All drug therapy was withdrawn during the York trial.

next couple of years she became aggressive and uncooperative. She kept on accusing the staff of trying to poison her, or to kill her in other ways.

She wrote many letters to the Duke of Edinburgh, and to Scotland Yard. She also telephoned the local police station about once a week to make accusations about the staff and other patients. All this gave ample evidence of paranoid delusions and the disordered thinking typical of a schizophrenic. She was not the slightest bit interested in her personal appearance or hygiene. She ate meals at irregular times, but even so, was very overweight. She smoked heavily, and went round picking up cigarette ends. Various treatments were tried, but without any marked effect.

In January 1978 she was put on penicillin V. In the first week she became less agressive, paranoid and hypochondriacal. She kept up a steady improvement until mid-May, when she was withdrawn from penicillin. By the third week after withdrawing the drug, she had once again become very aggressive, paranoid and uncooperative. In June 1979 she was put on a combination of penicillin and evening primrose oil. At first she complained of some nausea and headache, but these soon went away.

Within a few weeks, the delusions had disappeared. When she was shown her letters to the Duke of Edinburgh and Scotland Yard she was amused and said, 'I certainly would not write such things now.' She also stopped ringing up the local police station. She now writes sensible letters to her family, and has reasonable conversations with the staff. Her personal appearance and hygiene have improved beyond recognition, and she has even bought new clothes. She has cut down her smoking to about ten cigarettes a day; she eats at regular meal times, and has lost about 6kg in weight.

Case 2. A thirty-one-year-old man, with above average intelligence, but antisocial and introverted. When he was admitted to hospital in June 1966 he said he was hearing voices of God speaking to him, and claimed he was Jesus Christ. He was treated with drugs and ECT. Over the next five years he was admitted to hospital three times with acute schizophrenic relapses, and he became an in-patient in 1971. Since then, he had been on several different drug treatments, but his psychosis remained pretty much the same.

In April 1980 he was taken off all drugs, and instead put on evening primrose oil and penicillin V. At that time he was suffering from severe thought disorder; he could not speak spontaneously to other people, and he had a habit of standing in front of a mirror where he would laugh and talk to himself. His face had a wild, staring look on it. He often complained about hearing voices, and was aggressive to other patients. It was very hard to make him take part in any of the ward's activities.

Within the first month of this treatment he became more co-operative, and spoke more spontaneously. He stopped having hallucinations, and his thinking became clearer. Little by little, he became very co-operative on the ward, and was no longer easily upset by his fellow patients. After five months, his aggressive behaviour had completely gone, and he seemed almost normal.

Other treatment methods

It is not suggested that evening primrose oil and penicillin together is the only way of helping schizophrenics. Nor is it claimed that this approach is in any sense a cure. This regime is helpful against some of the 'negative' aspects of schizophrenia (e.g. emotional and social withdrawal and lack of feelings). However, current dopamine-blocking drugs are more effective against some of the 'positive' aspects of the disease (e.g. hallucinations and bizarre behaviour).

Cutting out wheat, milk, and foods containing arachidonic acid (meat, dairy products, peanuts) has also been of help in some patients. Some schizophrenics seem to be zinc-deficient and respond to zinc therapy. It could be that a combination of all these regimes may be the most effective treatment.

Further research

There is now no doubt that there is abnormal EFA and PG metabolism in schizophrenics. This means it might be possible to find a diagnostic test for schizophrenia. This is now being investigated at the Efamol Research Institute in Nova Scotia, Canada.

CAUTION: It has been found that patients with a special form of epilepsy known as temporal lobe epilepsy (who were mistakenly thought to be schizophrenics) got *worse* on high doses of evening primrose oil.

Lower doses (less than twelve capsules per day) did not affect temporal lobe epilepsy and no dose has been found to have adverse effects on any other form of epilepsy. However, evening primrose oil should be given cautiously to epileptic patients.

15. ALCOHOLISM

There is both good news and bad news about alcohol. The good news is that a little does you good. The bad news everyone knows already — a lot of alcohol does you harm. But the other bit of good news is that evening primrose oil can do a great deal to lessen the worst effects of heavy drinking.

A little of what you fancy does you good...

Recent research in North America and W. Europe has unanimously come to the conclusion that a little alcohol is good for you. The risk of death from heart disease is reduced in people who have a modest intake of alcohol — perhaps a glass or two of wine, or a beer, or one or two drinks of whisky a day. At this level of drinking, people tend to live longer than teetotallers, and much longer than people who drink more alcohol.

The relevance of this finding to evening primrose oil is that alcohol has an interesting effect on PGE1. At low levels, alcohol has a beneficial effect on PGE1. It seems that almost all the good things about alcohol — the feeling of being happy and just nicely intoxicated — may be due to an increase in PGE1.

A small amount of alcohol is known to have a beneficial effect on certain conditions. After all, drinking a hot toddy before going to bed has long been a traditional remedy for colds and flu. This may work because PGE1 is able to stimulate the body's immune system and help it to resist infection.[3,4,5]

... but you can have too much of a good thing

Too much drink has the opposite effect on the body. It robs it of Prostaglandin E1. PGE1 is made from a store of dihomo-gammalinolenic acid (DGLA) in the cells. With heavy drinking, these stores run very low. Then, especially when alcohol is withdrawn, the PGE1 levels fall far below normal with potentially catastrophic results. This dramatic fall in PGE1 may account for the hangovers, withdrawal symptoms and depression that so often go with heavy drinking.

Too much alcohol is a blocking agent. It gets in the way of the metabolic conversion process of cis-linoleic acid to gammalinolenic acid. So even though a heavy drinker may be eating plenty of foods with linoleic acid, they won't be getting through.[1,2,3] Because of this block, heavy drinkers are very likely to be low in PGE1. So alcohol uses up the stores of DGLA, and at the same time prevents those stores from being replenished from the usual diet.

The low levels of Prostaglandin E1 in alcoholics may have other serious consequences. This includes the risk of heart attacks and strokes, high blood pressure, a reduced ability to cope with infections, brain and nerve deterioration, and liver damage.

Mood and Prostaglandin E1

Most people drink socially because it makes them feel good. This may well be because, without knowing it, they are raising their levels of PGE1.[4] Studies on manic people, who feel abnormally euphoric, have found that they make more PGE1 than normal.[5] People who are feeling low often crave a drink because they know it will boost their spirits, at least for a time. It has been found that depressed people have lower than normal levels of PGE1. A small amount of drinking will raise their PGE1 levels, but a long bout will actually rob them of PGE1 and make them feel low all over again. This is the vicious circle of the alcoholic.

Evening primrose oil and the problems of alcohol

Hangovers. Evening primrose oil is highly effective in preventing hangovers. Doctors researching it have tried this treatment for themselves and found that four to six capsules straight after drinking and before going to bed greatly reduces the symptoms of hangover. It may well be that evening primrose oil is the best cure for hangovers since 'the hair of the dog'!

Withdrawal symptoms and post-drinking depression. Preliminary tests in humans have shown that evening primrose oil can make withdrawal from alcohol easier and can relieve post-drinking depression.[8]

Dr John Rotrosen and Dr David Sagarnick at New York University did similar tests on mice.[5] They made mice addicted to alcohol by giving them an alcohol-rich diet. They then took away the alcohol abruptly and over the next few hours there was a dramatic withdrawal syndrome, similar to what happens with human alcoholics. The doctors then injected PGE1 into the animals. This dramatically alleviated the withdrawal problems in the addicted mice. Tremor, irritability, over-excitability and convulsions were all reduced by about 50%.

Evening primrose oil had the same effect as PGE1 in preventing withdrawal symptoms. But when evening primrose oil was given with a drug which blocked its conversion to PGE1, the evening primrose oil failed to work.

Tolerance and addiction. People who crave drink need to drink more and more to achieve the same effect on their mood because they become tolerant to alcohol. This tolerance to high amounts of alcohol is a major factor in addiction.

In animal studies, tolerance could be prevented by giving evening primrose oil (*Efamol*) daily with alcohol. If evening primrose oil can prevent the development of tolerance, then maybe it can also help prevent addiction. Controlled studies of this in humans are now in progress.

Liver damage. A very recent study done at the Alcoholic Clinic at Craig Dunain Hospital in Inverness[8] showed that *Efamol* can go a long way in correcting liver damage due to alcohol.

Under consultant psychiatrist Dr Iain Glen, the clinic conducted a double-blind trial with about 100 patients. No one knew who was taking capsules of evening primrose oil, and who was taking identical capsules containing liquid paraffin.

The group who took the evening primrose oil (*Efamol* 500) did much better than the others. The results showed that evening primrose oil can improve liver function, reduce the demand for tranquillizers, improve brain function, and lower the incidence of hallucinations during the period of alcohol withdrawal. The liver seemed to benefit in particular; its biochemistry returned to normal much more rapidly among the patients taking evening primrose oil.

Dr Glen was working on the hypothesis that drink can seriously alter the body's membrane lipids. At the time when he delivered his paper to an International Conference on Pharmacological Treatments for Alcoholism in London in March 1983, he was quoted as saying: 'We used evening primrose oil because it contains a large amount of gamma-linolenic acid. These membrane changes can block the linoleic acid metabolism so, by giving alcoholics capsules of *Efamol*, we hoped to bypass this trouble.' Dr Glen said that *Efamol* is the first specific medicine to show promise in treating alcohol dependence.

Foetal alcohol syndrome. A wide range of foetal abnormalities can be produced by alcohol.[7] These include underweight and underlength at birth, with slow growth and failure to thrive even with special care. In humans, the babies tend to have unusually small heads, with defective development of some features, such as eyes set very wide apart. There can also be mental backwardness, behavioural problems, and extreme nervousness.

There is strong evidence to suggest that a block in the conversion of linoleic acid to gammalinolenic acid because of alcohol may be the cause of many of the alcohol-induced abnormalities. When evening primrose oil (*Efamol*) was given to laboratory animals with the alcohol, most of the abnormalities were prevented.[9]

CAUTION: Pregnant women, of course, should *not* drink.

16. CANCER

It's not such a crazy idea to suggest that evening primrose oil might be able to help cancer too, the most dreaded and dreadful of human diseases.

So far, all the research done on evening primrose oil and its effect on cancer cells has been done in test tubes in the laboratory, but even so, the exciting results point to promising future developments.

Recent research in South Africa[1] (originally published in the *South African Medical Journal*) showed that gammalinolenic acid, taken from evening primrose oil, reduced cancer cell growth by up to 70%. GLA was added to three different types of malignant cell, both human and mouse. The mouse cancer cells were inhibited, and the human cancer cells taken from the oesophagus were killed. This research showed that although GLA was toxic to malignant cells, it had no such effect on normal cells.

The Newsletter of the Northwest Academy of Preventive Medicine in the U.S.A.[2] published these comments, together with the findings: 'These data may have profound implications for the prevention and treatment of cancer. Whereas the usual method of fighting cancer is to destroy the malignant cells, GLA may be capable of actually reversing, or retarding, the malignant process. Of prime importance is the fact that GLA is a normal metabolite and is essentially non-toxic.'

How could evening primrose oil help?

Cancer cells multiply when the body's defense system isn't efficient enough to stop them. Evening primrose oil may help by converting into PGE1. And it is PGE1 which has such a crucial effect on the immune system, stimulating the T-cells and keeping the cell membranes healthy.

The immune system can only work effectively if it is given the right raw materials, and enough of them. As well as gammalinolenic acid, the body needs Vitamin C, Vitamin B_6, and zinc to make enough PGE1.

Laboratory tests have shown that it is this PGE1 which can have such a dramatic effect on cancer cells. When human and other cells are treated with cancer-causing agents such as radiation, certain chemicals, and cancer-causing viruses, they became transformed. This means that, in test tube conditions, they behave pretty much like malignant cells. They proliferate fast and develop a distorted function and structure. In the transformation process, the cells lose their ability to convert cis-linoleic acid to GLA, and therefore to make PGE1.[3,4,5] But when PGE1 or GLA is added to these transformed cells, they begin to behave normally.[5,6,7,8]

Cancer cells produce large amounts of 2 series PGs, and cannot make PGE1.[8] There is a possibility that malignant transformation may be produced by the loss of PGE1, and that the affected cells might return to normal when they are given enough PGE1.

Another fruitful line of enquiry is based on the hypothesis that in some forms of human cancer there is a loss of activity of the crucial enzyme, delta-6-desaturase, which is needed to convert linoleic acid to GLA. If this enzyme is out of action, then the body can't make enough PGE1.[5] Evening primrose oil, with gammalinolenic acid as its active ingredient, by-passes the delta-6-desaturase block which means that PGE1 can be made normally no matter how defective that enzyme might be.

If this hypothesis is correct, then ideally GLA should be used in combination with other co-factors which help the conversion process of cis-linoleic acid through to Prostaglandin E1*. There is a possibility that Vitamin C itself may have anti-cancer effects.[9] So trials are now in progress in several places to see what happens when Vitamin C and GLA are used as a combined treatment.

---

* Evening primrose oil is one of the nutritional therapies recommended by Bristol Cancer Help Centre, run by Dr Alec Forbes.

However, it must be stressed that there is no definitive information as yet on evening primrose oil and human cancer.

17. FURTHER RESEARCH

Evening primrose oil may turn out to be one of the most revolutionary nutritional finds of this century. But it is still very early days, and a great deal of research still needs to be done.

More trials are in progress or are planned for many of the conditions already listed in this book. But there are several new projects in the pipeline. Here are some of them:

- Pre-eclampsia in pregnancy.
- Infertility.
- Menstrual blood loss.
- Hepatic cancer.
- Essential tremor.
- Lithium toxicity.
- Depression.
- Anorexia Nervosa.
- Mental degeneration in the elderly.
- Terminal cancer.
- Diabetic skin lesions.
- Primary biliary cirrhosis.
- Skin diseases.
- Diabetes and diabetic eye disease.
- Migraine.
- Parkinson's Disease.
- Chronic diarrhoea.
- Cosmetic uses.

Side effects of evening primrose oil

Overall, the most common side effect from taking evening primrose oil is headache. This is more likely to happen with people who get headaches from alcohol. A way to avoid headaches is to take evening primrose oil capsules with food at meal times. If you have a tendency to get headaches, do not take the capsules last thing at night. (On the other hand, many people, especially those with PMS or migraine headaches, have reported improvement with evening primrose oil.)

Some people have felt mild nausea when they start taking evening primrose oil capsules, but this usually passes in a short time.

Some people say that they have softer stools. But most people who get this think of it as a bonus rather than a bad side effect. Some people have reported minor skin rashes, but these have tended to go away.

Evening primrose oil should be given cautiously to epileptics. See footnote on p. 83.

To prevent oxidation of the oil inside the body, evening primrose oil must be taken with vitamin E. (See appendix II).

Queries regarding research should be directed to:

Dr David Horrobin M.A., D. Phil., B.M., B.Ch.
President and Research Director
Efamol Research Inc.
Annapolis Valley Industrial Park
P O Box 818
Kentville
Nova Scotia
Canada B4N 4H8
Tel. area code 902 678-5534

APPENDIX I
USEFUL NAMES AND ADDRESSES

Chapter 5. Premenstrual Syndrome.
Women's Health Concern (WHC)
16 Seymour Street
London W1V 5WR
01-486-8653
Run by Joan Jenkins.

Chapter 10. Hyperactive Children.
Hyperative Children's Support Group
Secretary: Sally Bunday
59 Meadowside
Angmering
Sussex BN16 4RW
Tel. Rustington (09062) 70360

Chapter 13. Multiple Sclerosis.
Action for Research into Multiple Sclerosis (ARMS)
11 Dartmouth Street
London SW1H 9BL
Tel. 01-222 3224
Chairman: James Reilly

Chapter 14. Schizophrenia.
The Schizophrenia Association of Great Britain
Hon Sec: Mrs Gwynneth Hemmings
Tyr Twr, Llanfair Hall
Caernarfon LL55 1TT
Wales
Tel. Port Dinorwic (0248) 670379

Manufacturers of evening primrose oil in Britain

Efamol

Manufactured by:
Efamol Ltd.
Gable House
Woodbridge Meadows
Guildford
Surrey
Tel. (0483) 578060

Distributed by:
Britannia Health Products Ltd.
Lonsdale House
7-11 High Street
Reigate
Surrey
Tel. Reigate (74) 22256

Efamol 250 and *Efamol* 500 are available in chemists and health food shops.

Naudicelle

Manufactured by:
Bio-Oil Research Ltd.
Cecil House
Hightown
Crewe
Cheshire CW1 3BJ
Tel. Crewe (0270) 213094 and 213456

Naudicelle Plus is available direct from Bio-Oil by mail order.

Evening primrose oil is also available from:

Evening Primrose Oil
Co. Ltd.
17 Royal Crescent
Cheltenham
Gloucs. GL50 3DA
Tel. Cheltenham (0242)
39565

Cantassium Ltd.
225 Putney Bridge Rd.
London SW15 2PY

APPENDIX II
CAPSULE COMPOSITION

Efamol
Each *Efamol* 500 capsule contains 500mg of pure evening primrose oil, plus 10mg vitamin E.

Linoleic acid —	72%
Gammalinolenic acid, at least —	7%
Natural vitamin E —	10mg

Efamol 250 capsules contain:

Evening primrose oil —	250mg
Safflower oil —	200mg
Linseed oil —	50mg
Natural vitamin E —	10mg

Principal other fatty acids (approx):

Palmitic —	6%
Stearic —	2%
Oleic —	11%

Naudicelle Plus
Each *Naudicelle Plus* capsule contains 0.6ml of pure oil of evening primrose, plus marine oil. It provides approx. 72% essential fatty acids in the form:

Linoleic acid —	60%
Gammalinolenic acid —	7%
Eicosapentanoic acid —	2.5%
Vitamin E —	7.5IU

REFERENCES

Chapter 1. What is the Evening Primrose?
1. Erichson-Brown C. *Use of Plants for the Past 500 Years* (Breezy Creeks Press, Aurora, Ontario, 1980)
2. Unger, *Apothek Zeitgesellschoft* 1917, 32, 351
3. Heiduschka and Luft, *Archives of Pharmocology* 257 (1919), 33
4. Eibner, Widenmeyer, Schild. *Chemische Umschaurung* 34 (1927), 312
5. Riley J.P., *Journal of the Chemical Society* (1949), 2728-2731.
6. Hassam A.G., Sinclair A.J., Crawford M.A., 'The Incorporation of orally fed radioactive gammalinolenic acid and linoleic acid into the liver and brain lipids of suckling rats', *Lipids* 7 (1975), 417–420
7. Hassam A.G., Rivers J.P., Crawford M.A., 'Metabolism of gammalinolenic acid in essential fatty acid-deficient rats', *Journal of Nutrition* 4 (April, 1977)
8. Millar J.H.D., Zilkha K.J., Longman M.J.S., et al. *British Medical Journal* 1 (1973), 765
9. For all references to Field E.J., see pp. 107–108, refs. 6, 17, 18, 19, 21
10. Abdulla G.H., Hamadah, K., 'Effect of ADP on PGE formation in blood platelets from patients with depression, mania and schizophrenia', *British Journal of Psychiatry* 127 (1975), 591-5
11. John Williams, *Bio-Oil Research Ltd., Historical Background* (1977)

Chapter 2. A Unique Botanical Specimen
1. Stubbe W., Raven PH., 'A genetic contribution to the taxonomy of oenothera sect. Oenothera.' *Plant Systematics and Evolution* 133 (1979); 39, 59

Chapter 3. Essential Fatty Acids

1. Mead J.F., Fulco A.J., *The Unsaturated and Polyunsaturated Fatty Acids in Health and Disease* (C.C. Thomas, Springfield, 1976)

2. Gibson R.A., Kneebone G.M., 'Fatty acid composition of human colostrum and mature breast milk', *American Journal of Clinical Nutrition* 34 (1981), 252-7

3. Horrobin D.F., *Prostaglandins: Physiology, Pharmacology and Clinical Significance* (Eden Press, Montreal, 1978)

4. W.H.O./F.A.O., *Dietary fats and oils in human nutrition*. Report of an Expert Consultation (U.N. Food and Agriculture Organisation Rome, 1977)

5. Frankel T.L., Rivers J.P.W., 'The nutritional and metabolic impact of gammalinolenic acid on cats deprived of animal lipids'. *British Journal of Nutrition* 39 (1978), 227-31

6. Brenner R.R., 'Metabolism of endogenous substrates by microsomes', *Drug Metabolism Review* 6 (1977), 155-212

7. Pelutto R.O., Avala S., Brenner R.R., 'Metabolism of fatty acids of the linoleic acid series in testicles of diabetic rats', *American Journal of Physiology* 218 (1970), 669-73

8. Bailey J.M., 'Lipid metabolism in cultured cells', *Lipid Metabolism in Mammals II* (Ed. Snyder, F., Plenum Press, New York, 1977), 352

9. Horrobin D.F., 'A biochemical basis for alcoholism and alcohol-induced damage', *Medical Hypotheses* 6 (1980), 929-42

10. Kummerow F.A., 'Nutrition imbalance and angiotoxins as dietary risk factors in coronary heart disease', *American Journal of Clinical Nutrition* 32 (1979), 58-83

11. Holman R.T., Aaes-Jørgensen E., 'Effects of trans fatty acid isomers upon essential fatty acid deficiency in rats', *Proceedings of the Society for Experimental Biology and Medicine* 93 (1956), 175-9

12. Privett O.S., Phillips F., Shimasaki H., et al. 'Studies of the effects of trans fatty acids in the diet on lipid metabolism in essential fatty acid deficient rats', *American Journal of Clinical Nutrition* 30 (1977), 1009-17

13. Kinsella J.E., Hwang D.H., Yu P., Mai J., Shimp J., 'Prostaglandins and their precursors in tissues from rats

fed on trans-trans-linoleate', *Biochemical Journal* 184 (1979), 701-4

14. Anderson J.T., Grande F., Keys A., 'Effect on serum cholesterol in man of fatty acids produced by hydrogenation of corn oil', *Federal Proceedings* 20 (1961), 96

15. Thomas L.H., Jones P.R., Winter J.A., Smith H., 'Hydrogenated oils and fats: the presence of chemically modified fatty acids in human adipose tissue', *American Journal of Clinical Nutrition* 34 (1981), 877-86

16. Cook H.W., 'Incorporation, metabolism and positional distribution of trans-unsaturated fatty acids in developing and mature lungs', *Biochimica Biophysica Acta* 531 (1978), 245-56

17. Hsu C.M.L., Kummerow F.A., 'Influence of eluidate and erucate on heart mitochondria', *Lipids* 12 (1972), 486-94

18. Yu P., Mai J., Kinsella J.E., 'The effects of dietary 9-trans, 12-trans-octadecadienoate on composition and fatty acids of rat lungs', *Lipids* 15 (1980), 975-9

19. Enig M.G., 'Fatty acid composition of selected food items with emphasis on trans octadecenoate and trans octadecedienoate', (Thesis, University of Maryland, 1981)

20. Enig M.G., Munn R.J., Keeney M., 'Dietary fat and cancer trends', *Federal Proceedings* 37 (1978), 2215-20

21. Beare-Rogers J.L., Gray L.M., Hollywood R., 'The linoleic and trans fatty acids of margarines', *American Journal of Clinical Nutrition* 32 (1979), 1805-9

22. Coots R.H., 'Metabolism of geometric isomers of linoleic and oleic acids in the rat', *Nutrition Review* 23 (1929), 12-14

23. Burr G.O., Burr M.M., 'A new deficiency disease produced by the rigid exclusion of fat from the diet', *Journal of Biological Chemistry* 82 (1929), 345-367

Chapter 4. Prostaglandins

1. Huttner J.J., Gwebu E.T., Pangamala R.V., et al., 'Fatty acids and their prostaglandin derivatives: inhibitors of proliferation in aorta smooth muscle cells', *Science* 197 (1977), 189-91

2. Gemsa D., Seitz M., et al., 'The effects of phagocytosis,

dextran and cell damage on PGE1 sensitivity and PEG1 production of macrophages', *Journal Immunol* 120, 1187-94

3.   Lagarde M., Dechavanne M., et al., 'Basal level of platelet prostaglandins: PGE1 is more elevated than PGE2', *Prostaglandins* 17 (1975), 685-91

4.   Casey C.E., Meydani S., et al., 'Prostaglandins in human duodenal secretions', *Prostaglandins and Medicine* 4 (1980), 449-52

5.   Karim S.M.M., Sandler M., Williams E.D., 'Distribution of prostaglandins in human tissues', *British Journal of Pharmacology* 31 (1967), 340-4

6.   Perry D.L., Desiderio D.M., 'Endogenous levels of PGE1, PGE2, 190H-PGE1 and 190H-PGE2 in human seminal fluid by GC-MS-SIM', *Prostaglandins* 14 (1977), 745-52

7.   Jugdutt B.I., 'Prostaglandins in myocardial infarction with emphasis on myocardial preservation', *Prostaglandins and Medicine* (in press)

8.   Zurier R.B., Ballas M., 'Prostaglandin E suppression of adjuvant arthritis', *Arthritis and Rheumatism* 16 (1973), 251-6

9.   Rotrosen J., Mandis D., Segarnick D., et al., 'Ethanol and prostaglandin E1: biochemical and behavioural interactions', *Life Science* 26 (1980), 1867-76

10.   Johnson G.S., Friedman R.H., Pastan I., 'Morphological transformation of cells in tissue culture by dibutyryl cyclic AMP', *Proceedings of the National Academy of Science U.S.A.* 68 (1975), 425-9

11.   Puck T.T., 'Cyclic AMP, the microtubule/microfilament system and cancer', *Proceedings of the National Academy of Science U.S.A.* 74 (1977), 4491-5

12.   Stone K.J., Willis A.L., Hart M., et al., 'The metabolism of dihomo–gammalinolenic acid in man', *Lipids* 14 (1979), 174–80

13.   Horrobin D.F., *Prostaglandins: Physiology, Pharmacology and Clinical Significance* (Eden Press, Montreal, 1978)

14.   von Euler U.S., *Klinische Wochenschrift* 14 (1935), 1182

Chapters 5 and 6. Premenstrual Syndrome/Benign Breast Disease
1. Dalton K., *The Premenstrual Syndrome* (Heinemann, London, 1964)
2. Reid R.L., Yen S.S.C., 'Premenstrual Syndrome', *American Journal Obstetrics and Gynaecology* 139 (1981), 85-104
3. Kerr G.D., 'The management of the premenstrual syndrome', *Current Medical Research and Opinion* 4 suppl. 4 (1977), 29–34
4. Horrobin D.F., *Prolactin: Physiology, Pharmacology and Clinical Significance* (Eden Press, Montreal, 1973)
5. Cole E.N., Evered D., Horrobin D.F., et al., 'Is prolactin a fluid and electrolyte regulating hormone in man?' *Journal of Physiology* 252 (1975), 54
6. Benedek-Jaszmann L.J., Hearn-Sturtevant M.D., 'Premenstrual tension and functional infertility', *Lancet* 1 (1976), 1095–7
7. Halbreich V., Assael M., Ren-David M., et al., 'Serum prolactin in women with premenstrual syndrome', *Lancet* 2 (1976): 654-5
8. Horrobin D.F., 'Cellular basis of prolactin action', *Medical Hypotheses* 5 (1979), 599-620
9. Brush M.G., Taylor R.W., 'Gammalinolenic acid (*Efamol*) in the treatment of the premenstrual syndrome', (in press)
10. Pashby N.L., Mansel R.E., Preece P.E., Hughes L.F., Aspinall J. A., 'Clinical trial of evening primrose oil (*Efamol*) in mastalgia', British Surgical Research Society, Cardiff meeting, July 1981.
11. Dalton K., *The Premenstrual Syndrome and Progesterone Therapy* (Heinemann, London 1977)
12. Lever J. (with Brush M.G. and Haynes E.), *PMT - The Unrecognised Illness*, (Melbourne House, London, 1979)
13. Brush M.G., Watson M., 'Premenstrual syndrome, dysmenorrhoea and menstrual migraine', *Gynaecology in Nursing Practice* (Shorthouse M., Brush M.G., eds. Balliere, London, 1981), 60–71
14. Brush M.G., 'The possible mechanisms causing the premenstrual tension syndrome', *Current Medical Research and Opinion* 4, suppl. 4 (1977), 9-15

15.   Brush M.G., 'Endocrine and other biochemical factors in the aetiology of the premenstrual syndrome', *Current Medical Research and Opinion* 6, suppl. 5 (1979), 19-27
16.   Horrobin D.F., 'Cellular basis of prolactin action; involvement of cyclic nucleotides, polyamines, prostaglandins, stero thyroid hormones, na/K ATPase', *Medical Hypostheses* 5 (1979), 599-620

Chapter 7. Heart Disease, Vascular Disorders and High Blood Pressure

1.   Sinclair H., 'Dietary fats and coronary heart disease', *Lancet* 1 (1980), 414-5
2.   Broute-Stewart B., 'The effect of dietary fats on blood lipids and their relalation to ischaemic heart disease', *British Medical Bulletin* 14 (1958), 243-52
3.   Fleischman A.I., Watson P.B., Stier A., et al., 'Effect of increased dietary linoleate upon blood pressure, platelet function and serum lipids in hypertensive adult humans', *Preventive Medicine* 8 (1979), 163
4.   Hornstra G., Lewis B., Chair A., et al., 'Influence of dietary fat on platelet function in men', *Lancet* 1 (1973), 1155-7
5.   Siguel E., 'Triglycerides and HDL cholesterol', *New England Journal Medicine* 304 (1981), 424-5
6.   Kummerow F.A., 'Nutrition imbalance and angiotoxins as dietary risk factors in coronary heart disease', *American Journal of Clinical Nutrition* 32 (1979), 58-83
7.   Anderson J.T., Grande F., Keys A., 'Effect on serum cholesterol in man of fatty acids produced by hydrogenation of corn oil', *Federal Proceedings* 20 (1961), 96
8.   Bearc-Rogers J.L., Gray L.M., Hollywood R., 'The linoleic and trans fatty acids of margarines', *American Journal of Clinical Nutrition* 32 (1979), 1805-9
9.   Shparer A.G., Marr J.W., 'Fatty acids and ischaemic heart disease', *Lancet* 1 (1978), 1146-7
10.   Efamol Research Institute. Results on file
11.   Oster P., Arab L., Schellenberg B., et al., 'Blood pressure and adipose tissue linoleic acid', *Research in Experimental Medicine* (Berlin) 175 (1979), 287-91
12.   Ten Hoor F., 'Cardiovascular effects of dietary linoleic acid', *Nutrition Metabolism* 24, suppl. 1 (1980), 162-80

13. Castelli W.P., Doyle J.T., Gordon T., et al., 'Alcohol and blood lipids', *Lancet* 2 (1977), 153-7

14. Horrobin D.F., 'A new concept of lifestyle related cardiovascular disease: the importance of interactions between cholesterol, essential fatty acids, prostaglandin E1 and thromboxane A2', *Medical Hypotheses* 6 (1980), 785-800

15. Willis A.J., et al., 'Effects of essential fatty acids on platelet function in guinea pigs', *Progress in Lipid Research* (in press)

16. Sinclair H.M., 'Prevention of coronary heart disease. The role of essential fatty acids', *Postgraduate Medical Journal* 56 (1980), 579-84

17. Sim A.K., and McCraw A.P., 'The activity of gamma-linolenate and dihomo-gammalinolenate methyl esters in vitro and in vivo on blood platelet function in non-human primates and in man', *Thrombosis Research* 10 (1977), 385-397

18. Paper in *Lipids* (in press)

Chapter 8. Obesity
1. Trayhurn P., Fuller L., 'The development of obesity in genetically diabetic obese mice pair-fed with lean siblings', *Diabetologia*, 19 (1980), 148-153

2. James W.P.T., Trayhurn P., 'Thermogenesis and obesity', *British Medical Bulletin* 37 (1981), 43-48

3. DeLuise M., Blackburn G.L., Filier J.S., 'Reduced acitivity of the red cell sodium potassium pump in human obesity', *New England Journal of Medicine* 303 (1980), 1017-22

4. Avenall A., Leeds A.R., 'Sodium intake, inhibition of Na/K ATPase and obesity', *Lancet* 1 (1981), 836

5. Osler P., Arab L., Schellenburg B., et al., 'Blood pressure and adipose tissue linoleic acid', *Research in Experimental Medicine* (Berlin) 175 (1979), 287-91

6. Vaddadi K.S., Horrobin D.F., 'Weight loss produced by evening primrose oil administration in normal and schizophrenic individuals', *IRCS Journal of Medical Science* 7 (1979), 52

7. Evans et al., 'Weight loss and activation of erythrocyte

sodium transport during treatment with evening primrose oil', submitted for publication
8.   Efamol Research Institute. Results on file.
9.   Mir M.A., and others. Dept. of Medicine Metabolic and Diabetic Unit, University Hospital of Wales, Cardiff. Trial in progress.

Chapter 9. Eczema, Asthma, Allergies and Cystic Fibrosis

1.   Warner J.O., Norman A.P., Soothill J.F., 'Cystic fibrosis heterozygosity in the pathogenesis of allergy', *Lancet* 1 (1976), 990-991
2.   Rivers J.P.W., Hassam A.G., 'Defective essential fatty acid metabolism in cystic fibrosis', *Lancet* 2 (1975), 642-643
3.   Lloyd-Still J.D., Ahnson S.B., Holman R.T., 'Essential fatty acid status in cystic fibrosis and the effects of sunflower oil supplementation', *American Journal of Clinical Nutrition* 34 (1981), 1-7
4.   Wright S., Burton J.L., 'Effects of evening primrose oil (*Efamol*) on atopic eczema', *Lancet* (November 20 1982), 1120-1122
5.   Parish W.E., Champion R.H., 'Atopic dermatitis', *Recent Advances in Dermatology* (A. Rook, ed., Churchill Livingston, Edinburgh, 3rd edition 1973)
6.   Byron N.A., Timlin D.M., 'Immune status in atopic eczema: a survey', *British Journal of Dermatology* 100 (1979), 491-498
7.   Bach M.A., Bach J.F., 'Effects of prostaglandins and indomethacin on rosette-forming lymphocytes: interaction with thymic hormone'; *Prostaglandin Synthetase Inhibitors*, (Robinson H.J. and Vane J.R. eds., Raven Press, New York, 1974), 241-248
8.   Zurier R.B., Sayndoff D.M., Torrey S.B., et al., 'Prostaglandin E treatment in NZB/NZW mice', *Arthritis and Rheumatism* 20 (1977), 723-728
9.   Horrobin D.F., Manku M.S., Oka M., et al., 'The nutritional regulation of T-lymphocyte function', *Medical Hypotheses* 5 (1979), 969-985
10. Jackson P.G., Lessof M.H., Baker R.W.R., et al., 'Intestinal permeability in patients with eczema and food

allergy', *Lancet* 1 (1981), 1285-1286

11. Sinclair, H.M., 'Essential fatty acids and the skin', *British Medical Bulletin* 14 (1958), 258-262

12. Saarinen U.M., Katosaarir M., Backman A., et al., 'Prolonged breast feeding as prophylaxis for atopic disease', *Lancet* 11 (1979), 163-168

13. Midwinter R.E., Moore W.J., Soothill J.F., et al., 'Infant feeding and atopy', *Lancet* 1 (1982), 339

14. Gibson R.A., Kneebone G.M., 'Fatty acid composition of human colostrum and human breast milk', *American Journal of Clinical Nutrition* 34 (1981), 252-257

15. Crawford M.A., Hassam A.G., Rivers J.P.W., 'Essential fatty acid requirements in infancy', *American Journal of Clinical Nutrition* 31 (1978), 2181-2185

16. Dunbar L.M., Bailey J.M., 'Enzyme deletions and essential fatty acid metabolism in cultured cells', *Journal of Biological Chemistry* 250 (1975), 1152-1154

17. Brenner R.R., 'Nutritional and hormonal factors influencing desaturation of essential fatty acids', *Progress in Lipid Research* 20 (1982), 41-47

18. Brenner R.R., Peluffo R.O., 'Regulation of unsaturated fatty acid biosynthesis. 1. Effect of unsaturated fatty acid of 18 carbons on the microsomal desaturation of linoleic acid into gammalinolenic acid', *Biochimica Biophysica Acta* 176 (1969), 471-479

19. Sprecker H., 'Biochemistry of essential fatty acids', *Progress in Lipid Research* 20 (1982), 13-22

20. Horrobin D.F., 'A biochemical basis for alcoholism and alcohol-induced damage including the foetal alcohol syndrome and cirrhosis; interference with essential fatty acid and prostaglandin metabolism', *Medical Hypotheses* 6 (1980), 929-942

Chapter 10. Hyperactive Children

1. Feingold B.F., *Why Your Child Is Hyperactive* (Random House, New York, 1975)

2. *Lancet* Editorial, 'Feingold's regime for hyperkinesis', *Lancet* 2 (1979), 617-8

3. Swanson J.M., Kinsbourne M., 'Food dyes impair performance of hyperactive children in a laboratory

learning test', *Science* 207 (1980), 1485-7
4. Colquhoun V., Bunday S., 'A lack of essential fatty acids as a possible cause of hyperactivity in children', *Medical Hypotheses* 7 (1981), 681-6
5. Horrobin D.F., Manku M.S., Oka M., et al., 'The nutritional regulation of T-lymphocyte function', *Medical Hypotheses* 6 (1979), 969-85
6. Horrobin D.F., 'The regulation of prostaglandin biosynthesis: negative feedback mechanisms and the selective control of the formation of 1 and 2 series prostaglandins; relevance to inflammation and immunity', *Medical Hypotheses* 6 (1980), 687-709
7. Hyperactive Children's Support Group. Results on file.

Chapter 11. Skin, Hair, Eyes, Mouth and Nails
1. Sinclair H.M., 'Essential fatty acids and the skin', *British Medical Bulletin* 14 (1958), 258-261
2. Fleming C.R., Smith L.M., Hodges R.E., 'Essential fatty acid deficiency in adults receiving total parenteral nutrition', *American Journal Clinical Nutrition* 29 (1976), 976-83
3. Campbell A.C., MacEwan G.C., 'Treatment of brittle nails and dry eyes', *British Journal of Dermatology* (in press)
4. Horrobin D.F., Campbell A., 'Sjogren's syndrome and the Sicca syndrome; the role of Prostaglandin E1 deficiency. Treatment with essential fatty acids and Vitamin C', *Medical Hypotheses* 6 (1980), 225-232
5. Burr G.O., Burr M.M., 'A new deficiency disease produced by the rigid exclusion of fat from the diet', *Journal of Biological Chemistry* 82 (1929), 345-367

Chapter 12. Rheumatoid Arthritis
1. Horrobin D.F., 'The regulation of prostaglandin biosynthesis; negative feedback mechanisms and the selective control of formation 1 and 2 series prostaglandins: relevance to inflammation and immunity', *'Medical Hypotheses* 6 (1980), 687-709
2. Karim S.M.M., Sandler M., Williams E.D., 'Distribution of prostaglandins in human tissues', *British Journal of Pharmacology* 31 (1967), 340-4

3. Bach M.A., Bach J.F., 'Effects of prostaglandins on rosette forming lymphocytes', *Prostaglandin Synthetase Inhibitors* (eds. H.J. Robinson and J.R. Vane, Raven Press, New York, 1974), 241-48
4. Horrobin D.F., et al., 'The nutritional regulation of T-lyphocyte function', *Medical Hypotheses* 5 (1979), 969-85
5. Zurier R.B., Ballas M., 'Prostaglandin E suppression of adjuvant arthritis', *Arthritis and Rheumatism* 16 (1973), 251-6
6. Krakauer K., Torrey S.B., Zurier R.B., 'Prostaglandin E1 treatment of NZB/W mice', *Clinical Immunology Immunopathology* 11 (1978), 256-62
7. Kunkel S.L., Zurier R.B., 'Evening primrose oil and adjuvant arthritis', *Progress in Lipid Research* (in press)
8. McCormick J.N., Neill W.A., Sim A.K., 'Immunosuppressive effects of linoleic acid', *Lancet* 2 (1977), 508
9. Zurier R.B., Quagliata F., 'Effect of Prostaglandin E1 on adjuvant arthritis', *Nature* 234 (1971), 304-5

Chapter 13. Multiple Sclerosis
1. Sanders H., Thompson R.H.S., Wright H.P., Zilkha K.J., *Journal of Neurology, Neurosurgery, and Psychiatry* 31 (1968), 321
2. Millar J.H.D., Zilkha K.J., Longman M.J.S., et al., *British Medical Journal* 1 (1973), 765
3. Bates D., Fawcett P.R.W., Shaw D.A., Weightman D., *British Medical Journal* 2 (1978), 1390
4. Swank R.L., *Archives of Neurology* 23 (1970), 460
5. Swank R.L., *The Multiple Sclerosis Diet Book*, (Doubleday, New York, 1977)
6. Field E.J., Joyce G., *European Neurology Research* 17 (1978), 67
7. Mertin J., Stackpoole A., *Prostaglandins and Medicine* 1 (1978), 233
8. Seaman G.V.F., Swank R.L., Tamblyn C.H., Zukoski C.F., *Lancet* 1 (1979), 1138
9. Seaman G.V.F., Swank R.L., Zukoski C.F., *Lancet* 1 (1979), 1139
10. Bates D., Fawcett P.R.W., Shaw D.A., Weightman

D.A., *British Medical Journal* 2 (1978), 932

11.   Horrobin D.F., Manku M.S., Oka M., et al., *Medical Hypotheses* 5 (1979), 969

12.   Dunbar L.M., Bailey J.M., *Journal of Biological Chemistry* 250 (1975), 1152

13.   Efamol Research Institute. Results of file.

14.   Rosnowska M., Sobocinska Z., Cendrowski W., International Prostaglandin Meeting, Washington, (May 1979)

15.   Gul S., Smith A.D., Thompson R.H.S., *Journal of Neurology, Neurosurgery, and Psychiatry* 33 (1970), 506

16.   Mertin J., Shenton B.K., and Field E.J., *British Medical Journal* 1 (1973), 765

17.   Field E.J., Shenton B.K., *Acta Neurolgica Scandinavia* 52 (1975), 121

18.   Meyer-Rienecker H.J., Jenssen H.L., Kohler H., Field E.J., Shenton B.K., *Lancet* 11 (1976), 966

19.   Field E.J., Joyce G., *European Neurology Research* 17 (1978), 67

20.   'A statistical evaluation of *Naudicelle* as a dietary essential fatty acid supplement in multiple sclerosis patients', Bio-Oil Research Ltd.

21.   Field E.J., Shenton B.K., Joyce G., *British Medical Journal* 1 (1974), 412

22.   Hanlon K.B., Berkowitz G., Utermohlen V., Cornell University, U.S.A. Thesis

23.   Crawford M.A., Budowski P., Hassam A.G., *Proceedings of the Nutritional Society* 38 (1979), 373

Chapter 14. Schizophrenia

1.   Abdulla Y.H., Hamadah K., 'Effect of ADP on PGE formation in blood platelets from patients with depression mania and schizophrenia', *British Journal of Psychiatry* 127 (1975), 591-5

2.   Hitzemann J.J., Garver D.L., 'Abnormalities in membrane lipids associated with deficiencies in lithium counterflow', Society of Biological Psychiatry, New Orleans, (June 1981)

3.   Horrobin D.F., 'The regulation of prostaglandin biosynthesis; negative feedback mechanisms and the

selective control of formation of 1 and 2 series prostaglandins: relevance to inflammation and immunity', *Medical Hypotheses* 6 (1980), 687-709

4. Horrobin D.F., 'Prostaglandins and schizophrenia', *Lancet* 1 (1980), 706-7

5. Horrobin D.F., 'Schizophrenia: Reconciliation of the dopamine, prostaglandin and opioid concepts and the role of the pineal', *Lancet* (1979), 529-31

6. Chouinard G., Horrobin D.F., Annable L., 'An antipsychotic action of penicillin in schizophrenia', *IRCS Journal of Medical Science* 6 (1978), 187

7. Vaddadi K.S., 'Penicillin and essential fatty acid supplementation in schizophrenia', *Prostaglandins and Medicine* 2 (1979), 77-80

8. Vaddadi, K.S., 'Essential fatty acids in the treatment of schizophrenia', World Congress of Biological Psychiatry, Stockholm, (1981)

Chapter 15. Alcoholism

1. Reitz R.C., Wang L., Schilling R.J., et al., 'The nutritional and metabolic impact of gammalinolenic acid on cats deprived of animal lipid', *British Journal Nutrition* 39 (1978), 227-31

2. Holman R.T., Johnson S., 'Changes in essential fatty acid profile of serum lipids in human disease', *Progress in Lipid Research* (in press)

3. Horrobin D.F., 'A biochemical basis for alcoholism and alcohol-induced damage', *Medical Hypotheses* 6 (1980), 929-942

4. Horrobin D.F., Manku M.S., 'Possible role of Prostaglandin E1 in the affective disorders and in alcoholism', *British Medical Journal* 1 (1980), 1363-6

5. Rotrosen J., Mandio D., Segarnick D., et al., 'Ethanol and Prostaglandin E1: biochemical and behavioural interactions', *Life Science* 26 (1980), 1867-76

6. Wilson D.E., Engel J., Wong R., 'Prostaglandin E1 prevents alcohol-induced fatty liver', *Clinical Research* 21 (1973), 829

7. Jones K.L., Smith D.W., 'Recognition of the foetal alcohol syndrome in early infancy', *Lancet* 2 (1973), 999-1001

8.   Glen E., Glen I., Macdonnell L., Mackenzie J., 'Possible pharmacological approaches to the prevention and treatment of alcohol related CNS impairment. Results of a double blind trial of essential fatty acids', Highland Psychiatric Research Group, Craig Dunain Hospital, Inverness
9.   Work done by Professor T.V.N. Persaud, Head of Anatomy at the University of Manitoba Medical School

Chapter 16. Cancer
1.   Diepenaar et al., *South African Medical Journal* 62 (1982), 505; 683; Leary et al., *South African Medical Journal,* 62 (1982), 661
2.   Newsletter of Northwest Academy of Preventive Medicine, S. Africa Vol. 8, No. 1, (Jan 1983)
3.   Dunbar L.M., Bailey J.M., 'Enzyme deletions and essential fatty acid metabolism in cultured cells', *Journal of Biological Chemistry* 250 (1975), 1152-1154
4.   Bailey J.M., 'Lipid metabolism in cultured cells'; *Lipid Metabolism in Mammals II* (ed. Snyder F., Plenum Press, New York, 1977), 352
5.   Horrobin D.F., 'The reversibility of cancer: The relevance of cyclic AMP, calcium, essential fatty acids and Prostaglandin E1', *Medical Hypotheses* 6 (1980), 469-486
6.   Puck T.T., 'Cyclic AMP, the microtubule/microfilament system and cancer', *Proceedings of the National Academy of Sciences, U.S.A.* 74 (1977), 4491-4495
7.   Johnson G.S., Friedman R.H., Pastan I., 'Morphological transformation of cells in tissue culture by dibutyryl adenosine cyclic monophosphate', *Proceedings of National Academy of Science* U.S.A. 68 (1975), 425-429
8.   Horrobin D.F. et al., *Medical Hypotheses* 5 (1979), 969-985
9.   McCarthy D.M., May R.J., Maher M., Brennan M.F., 'Trace metal and essential fatty acid deficiency during total parenteral nutrition', *American Journal of Digestive Diseases* 23 (1978), 1009-1016

INDEX